ATOMIC PROCESSES
IN PLASMAS

AIP CONFERENCE PROCEEDINGS 257

ATOMIC PROCESSES IN PLASMAS

PORTLAND, ME 1991

EDITORS: EARL S. MARMAR
JAMES L. TERRY
MASSACHUSETTS INSTITUTE
OF TECHNOLOGY

American Institute of Physics   New York

Authorization to photocopy items for internal or personal use, beyond the free copying permitted under the 1978 U.S. Copyright Law (see statement below), is granted by the American Institute of Physics for users registered with the Copyright Clearance Center (CCC) Transactional Reporting Service, provided that the base fee of $2.00 per copy is paid directly to CCC, 27 Congress St., Salem, MA 01970. For those organizations that have been granted a photocopy license by CCC, a separate system of payment has been arranged. The fee code for users of the Transactional Reporting Service is: 0094-243X/87 $2.00.

© 1992 American Institute of Physics.

Individual readers of this volume and nonprofit libraries, acting for them, are permitted to make fair use of the material in it, such as copying an article for use in teaching or research. Permission is granted to quote from this volume in scientific work with the customary acknowledgment of the source. To reprint a figure, table, or other excerpt requires the consent of one of the original authors and notification to AIP. Republication or systematic or multiple reproduction of any material in this volume is permitted only under license from AIP. Address inquiries to Series Editor, AIP Conference Proceedings, AIP, 335 East 45th Street, New York, NY 10017-3483.

L.C. Catalog Card No. 92-71474
ISBN 0-88318-939-9
DOE CONF-9108105

Printed in the United States of America.

CONTENTS

Preface vii

COLLISION PROCESSES

Progress in Collisions of Multiply Charged Ions 3
   Ronald A. Phaneuf
Recombination of Highly Charged Ions with Free Electrons 15
   A. Müller, A. Frank, J. Haselbauer, G. Hofmann, J. Neumann, U. Pracht,
   E. Salzborn, S. Schennach, W. Spies, M. Stenke, O. Uwira, R. Völpel,
   M. Wagner, R. Becker, E. Jennewein, M. Kleinod, U. Pröbstel, R. A.
   Phaneuf, G. H. Dunn, E. M. Bernstein, N. Angert, and P. H. Mokler
Dielectronic Recombination Measurements of Highly-Charged
Helium-like and Neon-like Ions Using an Electron Beam Ion Trap 26
   M. B. Schneider, D. A. Knapp, P. Beiersdorfer, M. H. Chen, J. H. Scofield, C.
   L. Bennett, D. R. DeWitt, J. R. Henderson, P. Lee, M. A. Levine, R. E.
   Marrs, and D. Schneider

DENSE PLASMAS

A Review of Spectral Line Broadening Relevant to Hot Dense Plasmas 39
   R. W. Lee
X-Ray and Optical Diagnostics of Femtosecond Laser-Produced Plasmas 58
   P. Audebert, J. P. Geindre, J. C. Gauthier, R. Benattar, J. P. Chambaret,
   A. Mysyrowicz, and A. Antonetti
Recent Developments in the Super Transition Array Model for Spectral
Simulation of LTE Plasmas 68
   A. Bar-Shalom, J. Oreg, and W. H. Goldstein
Measurements and Models of the Opacity of Hot, Dense Plasma 78
   P. T. Springer, T. S. Perry, D. F. Fields, W. H. Goldstein, B. G. Wilson, and
   R. E. Stewart
The Inner-Shell $3d$–$4f$ Transitions in the Spectra of Highly
Ionized Heavy Elements 86
   P. Mandelbaum
Bound States and Ionization Kinetics in Dense Plasmas 97
   W. Ebeling, I. Leike, and U. Leonhardt

TOKAMAKS

Atomic Processes Relevant to Neutral Beam Based Tokamak Diagnostics 111
   H. P. Summers, M. von Hellermann, P. Breger, J. Frieling, L. D. Horton,
   R. Konig, W. Mandl, H. Morsi, R. Wolf, F. J. de Heer, R. Hoekstra, and
   W. Fritsch
Polarization Spectroscopy of Tokamak Plasmas 121
   Dariusz Wraoblewski

X-Ray Measurements from the JET and ASDEX Tokamaks 131
    U. Schumacher, R. Barnsley, G. Fußmann, K. Asmussen, C. C. Chu, and
    G. Janeschitz

Studies of Heating and Impurity Transport in the Plasma Boundary
of Tokamaks 144
    G. M. McCracken and U. Samm

PLASMA PROCESSING

Diagnostic Measurements in rf Plasmas for Materials Processing 157
    J. R. Roberts, J. K. Olthoff, M. A. Sobolewski, R. J. Van Brunt, J. R.
    Whetstone, and S. Djurović

ASTROPHYSICS

The Variability of Elemental Abundances in the Upper Solar Atmosphere 171
    Uri Feldman

The Fe *L*-Shell Spectrum in Compact Astrophysical X-Ray Sources 181
    D. A. Liedahl, S. M. Kahn, W. H. Goldstein, and A. L. Osterheld

Author Index 191

NOTE: The author who presented the talk is designated by *italic* type.

PREFACE

This volume contains papers based on 16 of the 25 invited talks presented at the Eighth American Physical Society Topical Conference on Atomic Processes in Plasmas which was held in Portland, Maine, 25–29 August 1991. The conference was attended by 175 scientists from around the world. The 1991 conference continued the tradition of providing a forum for researchers in the many areas of overlap between atomic and plasma physics. The previous conference in this series was held in Gaithersburg, Maryland, 2–5 October, 1989 (AIP Conference Proceedings No. 206).

The book begins with a review article summarizing the progress of the last two years in the experimental study of collision processes involving multiply charged ions. This is followed by two papers which examine various aspects of recombination processes.

Six papers deal with atomic processes in very high density plasmas. The subject of line broadening is covered in the first of these, with emphasis on theoretical progress. The advent of ultrashort pulse length lasers has led to an explosion of experimental work with irradiances in the 10^{17} W/cm^2 range, and some recent experimental results in this area are presented in the next paper. The third monograph examines recent progress in the use of the super transition array model for simulation of the spectra observed from these plasmas which can be very hot while approaching solid densities. Opacity of high density, high Z plasmas is important to a range of disciplines, from stellar astrophysics to inertial confinement fusion and x-ray lasers. The fourth paper of this section examines models for opacity in laser produced plasmas and presents comparisons with experimental observations. Emission from hot, dense, high Z plasmas is often dominated by inner-shell transitions in partially stripped ions, and the fifth paper deals with the theoretical and numerical analysis of such spectra. The state of theoretical understanding of the effects of extremely high densities on the bound levels of hydrogen and low Z atomic systems is presented in the last paper in this section.

Atomic processes have long played a crucial role in our ability to diagnose high temperature plasmas. In the case of magnetically confined fusion plasmas, a number of new techniques have been developed over the last few years, and 4 papers deal with applications to tokamak diagnostics. Many of the new techniques rely on the interaction of high energy neutral particles, which are introduced as a beam, with the plasma ions, and much of the related atomic physics is reviewed in one paper. A second paper examines various techniques which utilize the fact that the presence of the strong magnetic field causes line radiation to be polarized, which in turn allows for measurements of the magnetic field structure internal to the plasma. Since tokamak plasmas have electron temperatures in the 1 to 10 keV range, emission both continuum and line, is important in the soft x-ray range; experimental measurements are presented in a paper which emphasizes both the plasma parameter diagnostic aspects, as well as the possibilities for characterization of atomic processes and rate coefficient measurements. The edge plasma, including impurity generation and particle transport aspects, can also be investigated through spectroscopic techniques, and the fourth tokamak paper covers these topics.

Low temperature and density plasmas are widely used in the semiconductor manufacturing industry for materials processing. Measurements of the parameters of these plasmas are important for ensuring the success of the process, and spectroscopic approaches to this problem are presented in one article.

Two papers deal specifically with atomic processes as applied to astrophysical plasmas. The first deals with measurements and modeling of elemental abundances in the upper solar atmosphere as deduced from emission spectroscopy. The final article deals with L-shell emission from partially ionized iron found in compact astrophysical x-ray sources.

We are grateful to the Massachusetts Institute of Technology, the MIT Plasma Fusion Center, and the Office of Fusion Energy, U.S. Department of Energy, for providing partial support for the Conference and for the publication of these proceedings.

Earl S. Marmar
James L. Terry
Cambridge, Massachusetts
February 1992

COLLISION PROCESSES

PROGRESS IN COLLISIONS OF MULTIPLY CHARGED IONS

Ronald A. Phaneuf
Oak Ridge National Laboratory,* Oak Ridge, TN 37831-6372

ABSTRACT

The increasing power and availability of supercomputers during the last decade led to significant progress in the theory of multicharged ion interactions. However, important tests of many theoretical predictions were lacking, and have become possible only quite recently as new capabilities have been realized in the laboratory. This paper broadly surveys some of these experimental developments, and their impact on our understanding of collisional interactions of multicharged ions. The scope is limited to measurements made with monoenergetic beams.

INTRODUCTION

Collisional interactions of multiply charged ions are in general dominated by enhanced Coulomb or polarization forces, and/or by the appreciable neutralization energy that such ions carry into the interaction. This neutralization energy is equal to the sum of the binding energies of the q electrons that have been removed from an atom to produce an ion of net positive charge q. This may be thought of as a reservoir of stored electronic potential energy for certain classes of interactions. In experimental studies with low-energy ion beams, these q electrons are removed from the atom in an ion source prior to acceleration.

The large Coulomb forces between highly charged ions and other charged particles dominate long-range interactions, modify particle trajectories, and produce large cross sections for some processes. The large neutralization energy of the ion can overshadow kinetic effects in slow collisions. For multiply charged ions, the reduced screening of the nucleus due to missing electrons increases the binding energy of the outer valence electrons, but has a lesser effect on inner-shell electrons. Thus processes involving inner-shell electrons may become relatively more important as the ionic charge increases. For example, in electron-impact ionization of multiply charged ions, inner-shell excitation-autoionization can dominate direct outer-shell ionization by an order of magnitude or more in some cases.

Polarization forces due to ion-induced dipole interactions are also enhanced in multicharged ion interactions with neutral particles, since the attractive potential is proportional to the square of the ionic charge. This interaction is relatively weak and short-range compared to the Coulomb interaction, but can play a major role in multicharged ion-neutral collisions at electron-volt energies and below.

---

*Managed by Martin Marietta Energy Systems, Inc., for the U.S. Department of Energy under Contract No. DE-AC05-84OR21400.

The neutralization of multiply charged ions at solid surfaces has become a topic of intense worldwide activity during the past two years. This process is believed to preferentially populate outer shells of the ion, leading to the production of exotic "hollow" or "superexcited" atoms, whose relaxation has been studied by both photon and electron spectroscopy. This is one example in which theory currently lags experiment, but this advantage may be short-lived as more theorists are attracted to this new inter-disciplinary area.

Finally, the atomic structure of highly charged ions also provides perhaps the most critical testing ground for quantum electrodynamics, since relativistic effects and radiative corrections scale as Z^4. High-Z ions with simple electronic structures (e.g., H, He, Li or Na-like) provide particularly good cases for evaluating theory.

ADVANCED SOURCES OF MULTIPLY CHARGED IONS FOR RESEARCH

Our ability to study interactions of multiply charged ions in the laboratory is limited to a large extent by our ability to produce them under well-defined conditions. Powerful new sources of multiply charged ions have recently become available for the study of atomic interactions in many laboratories throughout the world, and have had a major impact on research in this area.

In the electron-cyclotron-resonance (ECR) ion source,[1] electrons confined by external magnetic fields are resonantly heated by absorption of microwave power, producing multicharged ions by successive ionization of atoms. This source produces intense, continuous ion beams of high charge which are accelerated to form beams for experiments. Improved performance in terms of higher beam intensity and charge states has recently been demonstrated at higher microwave frequencies and magnetic fields.[2] The technology of the ECR source is relatively straightforward, and many are currently in use throughout the world for atomic physics research. Without question, the ECR ion source has had the most widespread impact of any experimental device on the study of low energy collisions of multiply charged ions.

In the electron-beam ion source[3] (EBIS), multicharged ions are created by and trapped in the space charge well of an intense, high-energy, magnetically confined electron beam. An electrostatic drift-tube structure traps the ions axially, and is pulsed periodically to expel ions for acceleration. This source is capable of producing pulsed ion beams of very high charge for experiments. A novel feature is that the monoenergetic electron beam also permits electron-ion collision studies to be performed within the ion source itself.[4] However, the technology of the EBIS demanding, and only a relatively small number are being used for atomic physics research at the present time.

Sophisticated new experiments have been devised to take advantage of the ion beams from ECR and EBIS sources to probe interactions of multicharged ions with electrons and atoms. The techniques employed involve crossed and merged beams of multicharged ions with electrons and neutral atoms, and utilize charge analysis, energy-loss or ejected-electron spectroscopy to deduce quantitative data on collision processes such as ionization, excitation and electron capture.

Collisions of electrons with very highly ionized ions ($q \leq 80$) are now readily accessible to quantitative study using X-ray spectroscopy in an electron-beam ion trap[5] (EBIT). The EBIT is a compact version of the EBIS which employs an electron beam of very high energy, and is capable of producing and trapping ions of extremely high charge. A planned "super EBIT" will employ electron beams with energies as high as 250 keV. Ion beams have now also been successfully extracted from an EBIT for further collision studies.[6]

Another significant advance has been the application of ring accelerator technology to the storage of highly charged heavy-ion beams.[7] This has been made possible by effective ion-beam cooling obtained by merging an intense, monoenergetic electron beam with an ion beam circulating in a synchrotron ring at the same speed. The dramatically enhanced beam luminosity due to ion recirculation (by as much as a factor of 10^6) opens up a whole new spectrum of research possibilities. A number of heavy-ion storage-ring facilities have recently or will soon come on-line in Europe. Their raw power for high-resolution studies of electron-ion recombination has already been demonstrated by using the cooler electron beams part of the time to perform collision studies.[8] This same electron-cooler beam technology has also been successfully adapted to single-pass merged-beams experiments on conventional ion-beam accelerators in several laboratories.[9,10] These long-awaited high-resolution experiments have substantiated electron-ion recombination theory for low-Z multicharged ions, and planned measurements on higher-Z ions will provide more critical tests of theory.

The following sections document some experimental results obtained using these powerful new devices, and reported since the last meeting of this series in 1989. It should be borne in mind that there has been vigorous activity in this field, and only a few representative examples have been selected.

PROGRESS IN ELECTRON-IMPACT IONIZATION

An illustrative example to demonstrate the impact of multicharged ion source development on the study of electron-ion collisions is the electron-impact ionization of a simple Li-like (3-electron) ion, O^{5+}. The first crossed-beams measurements on this ion were reported in 1979 by Crandall et al.[11] using an O^{5+} ion beam of 10-40 nA produced in a Penning source at ORNL. These were the first cross-section measurements for ionization of a multicharged ion which clearly showed a contribution due to inner-shell excitation-autoionization superimposed on the direct ionization "background." The contribution of this indirect ionization process:

$$e + O^{5+}(1s^22s) \rightarrow e + O^{5+}(1s2snl) \rightarrow e + O^{6+}(1s^2) + e$$

to the total cross section was measured to be 20±10% near the threshold for the process at 550 eV, with the uncertainty due mainly to counting statistics on the ionization signal. The experiment was repeated by Crandall et al.[12] in 1984 using an improved crossed-beams apparatus and an O^{5+} ion beam of 300 nA from the newly-developed ORNL-ECR ion source operating in a preliminary mode at a microwave frequency of 2.45 GHz. In these new experiments, the 1s-nl excitation-autoionization contribution was measured to be 14±2%, providing one of the first definitive confirmations of theory for the inner-shell excitation cross section.

A year later, when the ORNL-ECR source became operational at its design frequency of 10.6 GHz, the O^{5+} ionization measurement was once again repeated by Rinn et al.[13] using a still more intense beam of O^{5+} ions. In these measurements, the statistical error bars of the measurements were reduced to less than 0.5%, and individual steps due to excitation-autoionization via the $1s2s^2$ and 1s2s2p levels were clearly visible for the first time. In this experiment, a search was also made for the predicted contribution of a more exotic process to the ionization cross section, the so-called resonant excitation auto-double ionization (READI) process.[14,15] The first step in this ionization mechanism is actually a capture of the incident electron into an autoionizing level, followed by the simultaneous ejection of two Auger electrons, for example:

$$e + O^{5+}(1s^2 2s) \rightarrow O^{4+}(1s 2s^2 2p) \rightarrow O^{6+}(1s^2) + 2e.$$

The most likely decay mechanism for the triply excited O^{4+} intermediate state is single autoionization, which produces only resonances in the elastic scattering cross section, since further autoionization is not possible for this level. This resonant process can only contribute to the ionization cross section via the relatively unlikely auto-double ionization process, for which the branching ratio is difficult to calculate, but is predicted to be extremely small. A detailed search was made in the 430-455 eV energy region, where these resonances are predicted to occur. While the data were suggestive, no clear evidence for the process could be claimed, due in part to the 2 eV energy spread in the electron beam.

The next chapter in the O^{5+} ionization story was written in 1990 in Giessen, Germany, using a newly-developed ECR ion source. The group led by A. Müller applied a rapid energy-scanning technique and a high-intensity space-charge-neutralized electron beam to the problem, and were successful in clearly demonstrating the READI process for the first time, as well as a rich resonance structure in the excitation-autoionization cross section due to the resonant excitation double ionization (REDA) process.[16] The latter process can occur at energies above the 1s-2s excitation-autoionization threshold at 550 eV, where decay of an intermediate triply excited O^{4+} state by two sequential single autoionization processes becomes possible. An example of the REDA process is as follows:

$$e + O^{5+}(1s^2 2s) \rightarrow O^{4+}(1s 2s 3lnl') \rightarrow O^{5+}(1s 2s^2) + e \rightarrow O^{6+}(1s^2) + 2e.$$

The Giessen measurements for O^{5+}, which have statistical uncertainties as low as 0.1%, are shown in Fig. 1. These data provide a definitive test of the theory for inner-shell excitation, as well as for the role of REDA resonances in a simple ion.

A comparison of similarly precise measurements[16] for ionization of Li-like C^{3+} with a recent close-coupling calculation[17] shows excellent agreement with the observed excitation-autoionization and REDA features in the cross section, providing the most stringent test of excitation theory to date. The measured READI contributions are also consistent with theoretical predictions[15] for the auto-double ionization process, although the latter are less accurate for this process.

Yet another chapter in the continuing O^{5+} ionization story is in progress, as reported by A. Müller at this conference.[10] An improved ECR ion source in Giessen is permitting such resonance structures to be measured in still finer detail! Such detailed studies of the indirect ionization process in simple few-electron ions are important in furthering our basic understanding of electron-ion interactions.

The presence of an inner electronic shell is a prerequisite for indirect ionization to occur, and Li-like ions are the simplest systems that satisfy this criterion, having only two tightly-bound 1s electrons in the inner shell. It is therefore no surprise that indirect contributions to the ionization cross section for Li-like ions are relatively small. However, this is not the case for ions with inner shells containing more, less-tightly-bound inner-shell electrons, where indirect ionization mechanisms may dominate the total ionization

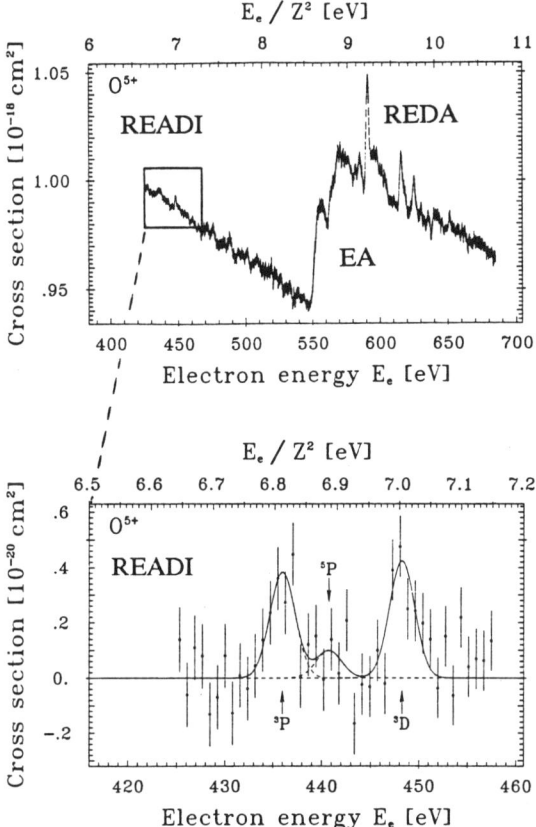

Fig. 1. Measured cross sections for electron-impact ionization of Li-like O^{5+} by Hoffman et al.,[16] showing contributions due to 1s-nl excitation-autoionization (EA), beginning at 550 eV in upper figure. Resonance features due to the REDA process are also evident in the 550-650 eV ranges. Lower figure shows resonance features due to READI process in the 430-460 eV range (box in upper figure). Figure is taken from Ref. 16.

cross section.[18] A case in point is ionization of Na-like Fe^{15+} ($2p^63s$) by electron impact. Figure 2 shows a comparison of experimental cross-section measurements of Gregory et al.[19] with theoretical calculations of Chen et al.[20] At energies above 800 eV, the contribution due to $2p^53snl$ excitation-autoionization is 4 times the direct ionization, and the REDA contributions ($2p^53snln'l'$) are comparable in magnitude to the direct ionization cross section. This calculation represents by far the most comprehensive treatment of electron-impact ionization to date, accounting for the excitation and decay of more than 10,000 different autoionizing levels, and requiring a tremendous computational effort. Unfortunately, the statistical precision of the experiment is insufficient to map out the resonance features in this case, but further measurements on Fe^{15+} are planned when an upgraded ECR ion source becomes available at ORNL.

PROGRESS IN ELECTRON-IMPACT EXCITATION

Much experimental data has been obtained on inner-shell excitation processes via their signatures in the ionization cross section, but far less experimental data is available for outer-shell excitation processes. All existing absolute excitation cross-section

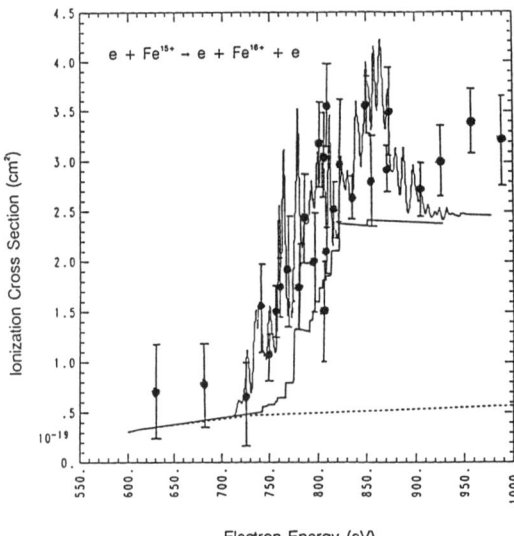

Fig. 2. Comparison of experimental cross-section measurements of Gregory et al.[19] with theoretical calculations of Chen et al.[20] for electron-impact ionization of Na-like Fe^{15+}. The dashed curve shows the direct ionization cross section; the middle solid curve includes the 2p-nl excitation-autoionization contributions, and the upper curve also includes contributions due to the REDA process. Figure is taken from Ref. 20.

measurements for multiply charged ions have been based on the crossed-beams approach and on absolute intensity measurements of the radiation emitted as the excited states decay.[18] Such measurements are severely limited by low photon intensities and detection efficiencies (~10^{-4}), difficulties with absolute optical calibration, as well as low electron energy resolution (~2eV). Cascade contributions from excitation of higher levels may also contribute to the measured photon emission, complicating such measurements.

A different experimental approach[21] which circumvents these limitations has been developed by the group of G.H. Dunn at JILA for use in conjunction with the ORNL-ECR multicharged ion source. The technique, illustrated in Fig. 3, involves merging fast beams of electrons and multiply charged ions in a uniform axial magnetic field, and detecting electron-impact excitation events via electron energy-loss spectroscopy. Trochoidal (E x B) analyzers are used to merge and demerge the beams. The magnetic field ensures complete collection of inelastically scattered electrons, which are directed onto a position-sensitive detector. The first absolute cross-section measurements on a multiply charged ion with this apparatus were reported recently[21] for 3s-3p excitation of Si^{3+}. This system was chosen because cross-section calculations for Na-like ions are expected to be of high reliability, and thus a comparison of experiment and theory serves to test the validity of both. The observed experimental energy resolution of 0.2 eV is an order of magnitude better than what has previously been achieved for excitation of multiply charged ions. Although the energy-loss method employed here is applicable only at energies near the excitation threshold, this is the energy range where the cross section and resonance effects are usually largest, and theory is most critically tested. This energy region also dominates the excitation rate in most plasma applications.

Another approach to the measurement of cross sections for electron-impact excitation of multiply charged ions has been developed by Huber et al.[22] at the LAGRIPPA ECR ion source in Grenoble. Energy- and angle-resolved measurements of elastically and inelastically scattered electrons are performed in a crossed-beams geometry. Absolute differential cross-section measurements have recently been reported for 3s-3p excitation of Na-like Ar^{7+} at scattering angles between 10° and 27°. Such measurements

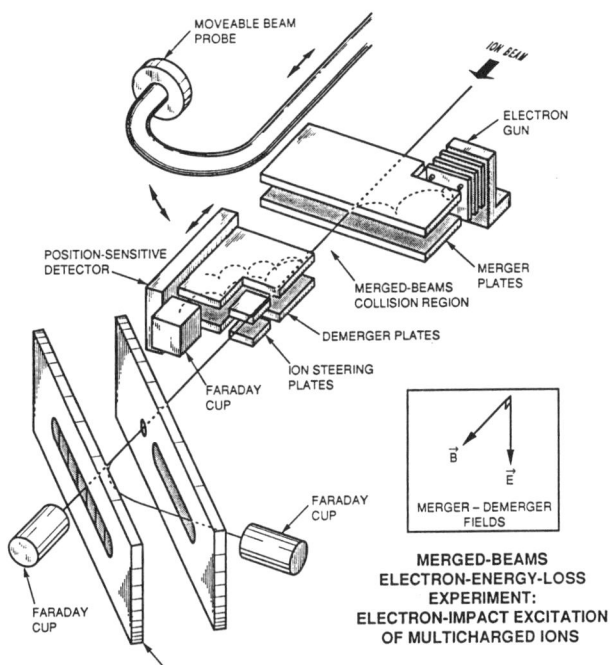

Fig. 3. JILA merged-beams electron-energy-loss apparatus for cross-section measurements of electron-impact excitation of multiply charged ions near threshold.

are complementary to total excitation cross-section measurements, and provide stringent tests of electron scattering theory.

PROGRESS IN ELECTRON-ION RECOMBINATION

Significant progress has been made during the past two years in high-resolution measurements of cross sections for dielectronic and radiative recombination. The successful application of several different technologies to this problem has provided long-awaited definitive tests (and verifications) of theoretical predictions for these processes for ions with simple electronic structure. Three of these were reported in *Physical Review Letters* during February 1990.

The first in this series was a measurement of the dielectronic recombination cross section for He-like Ar^{16+} by Ali et al.[4] at Kansas State University, using a newly commissioned EBIS. They measured the ion source yields of Ar^{16+} and Ar^{15+} as a function of electron beam energy in the 2.0-3.5 keV range, and determined the n=1 → n=2 dielectronic recombination cross section from their ratio, as shown in Fig. 4. Individual 1s2lnl' (n≤4) resonances were resolved in their measurements, which were placed on an absolute scale by normalization to theory for electron-impact ionization of Ar^{16+}. Reasonable agreement with theory was obtained when the calculated cross section was convoluted with an experimental electron energy resolution of 61 eV, although the theory appears to overestimate the contributions of the higher-n resonances.

The first atomic physics cross-section measurements in a heavy-ion storage ring were reported in *Physical Review Letters* that same month by Kilgus et al.[8] They used the merged electron-cooler beam in the Test Storage Ring at Heidelberg in Germany to

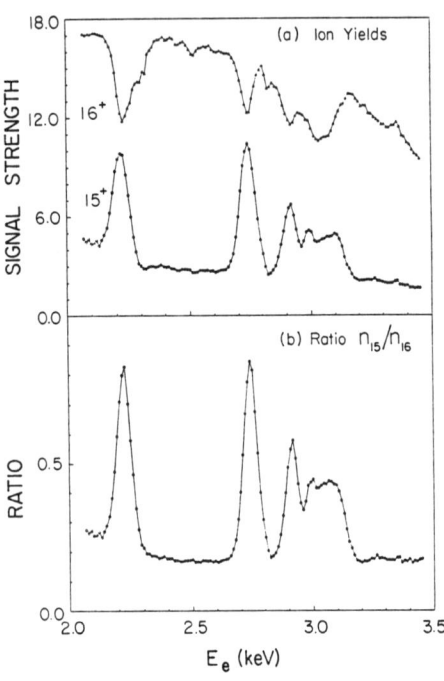

Fig. 4. Measured yields of Ar^{16+} and Ar^{15+}, and their ratio, as a function of electron-beam energy in the Kansas State EBIS, from Ref. 4. The ratio is proportional to the dielectronic recombination cross section.

measure absolute cross sections for dielectronic recombination of H-like O^{7+} in the 450-850 eV energy range. The laboratory energy of the O^{7+} beam in the ring was 143 MeV. With a center-of-mass energy resolution of 2 eV, they were able to resolve individual $2l2l'$ O^{6+} resonances as well as $2lnl'$ resonances with $n \geq 3$. The latter account for more than 90% of the total dielectronic recombination cross section. The measurements are in reasonable agreement with theoretical calculations of Griffin and Pindzola,[23] both with respect to the energies of the doubly-excited states, and the relative cross sections. They predict that by using separate cooling and target electron beams, the energy resolution could be improved by more than an order of magnitude. In that same issue, Andersen et al.[9] reported absolute rate coefficient measurements for radiative recombination of electrons with C^{6+} ions by merging an similarly intense, high-quality electron beam with a 24-MeV C^{6+} beam from a tandem accelerator. The relative energy range of these measurements was 0-1 eV, with an estimated resolution of 0.15 eV. Their data agree well with theoretical predictions for the process.

Measurements of X rays emitted as a function of the electron-beam energy in EBIT have permitted dielectronic recombination cross sections to be determined[24] for ions in such extremely high charge states such as Ne-like Xe^{44+}, Au^{69+} and Th^{80+}. The data analysis is complicated somewhat by the superposition of resonances of nearby charge states, but since the Ne-like stages dominate, it is possible to fit such data to predictions of Dirac Fock theory in order to extract individual cross sections, which are placed on an absolute scale by normalization to those for radiative recombination. The electron energy resolution of such measurements is of the order of 50 eV. The EBIT has also been applied to precision X-ray spectroscopy of very highly ionized systems,[25] such as Na-like Pt^{67+}, providing sensitive tests of QED theory.

PROGRESS IN ION-ATOM COLLISIONS

With the successful application of optical and translational-energy spectroscopic methods, much progress has been made in the quantitative study of state-selective electron-capture collisions of multiply charged ions during the last several years.[26] High-resolution Auger-electron spectroscopy has also been used as a sensitive probe of the double-capture process, which is of intrinsic interest in understanding ion-atom collision mechanisms, but of lesser direct importance in plasma applications.

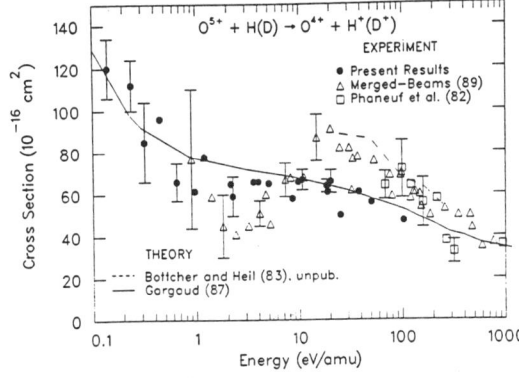

Fig. 5. Comparison of merged-beams electron-capture cross-section measurements with other measurements and theoretical calculations for O^{5+} + H collisions, from Ref. 28.

As noted earlier, the ion-induced-dipole (polarization) interaction between a highly charged ion and a neutral particle can play an important role in collision dynamics at low energies, such as those which prevail in the edge or scrape-off plasma of magnetic fusion devices. A new experimental approach[27] has been developed in which intense multicharged ion beams from the ORNL-ECR ion source are merged with a 5-10 keV beam of ground-state H atoms formed by laser photodetachment of a H⁻ beam. By varying the relative energies of the beams, absolute electron-capture cross sections have been measured at center-of-mass collision energies in the 0.1 - 2000 eV/amu range. Representative data[28] are shown in Fig. 5 for O^{5+} + H collisions. The rise in the cross section at energies below 2 eV/amu is due in part to the opening of the 4p capture channel, and in part to trajectory effects resulting from the attractive potential in the initial state. For more highly charged ions, such effects are expected to occur at higher energies, and resonances[29] may also appear in the total cross section due to population of quasi-bound levels in the attractive potential well in the initial state.

MULTICHARGED ION-SURFACE INTERACTIONS

A remarkable and unexpected result of recent research on the interaction of slow, highly charged ions with solid surfaces has been the extremely rapid neutralization which is observed to occur. Our understanding of such processes follows from the pioneering work of Hagstrum,[30] who first proposed the so-called resonance-neutralization model nearly 40 years ago, and from work in the 1970's by Arifov, Parilis and co-workers.[31] The process is illustrated in Fig. 6. As a highly charged bare (or nearly bare) ion approaches a metal surface, the potential barrier between the ion and the surface is lowered. At some critical distance, electrons flow from the valence band of the metal into empty excited levels of the ion, forming so-called "hollow" or "super-excited" atoms. Of course,

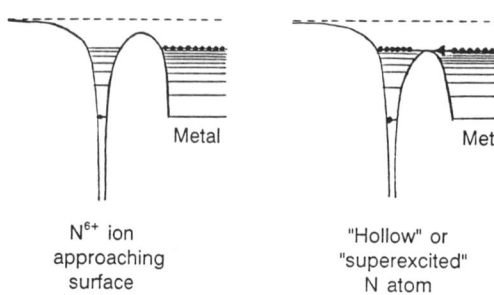

Fig. 6. Formation of a "hollow" or "super-excited" N atom by the resonance neutralization of a N^{6+} ion at a metal surface.

quantum-mechanical tunnelling through the barrier is possible at larger distances. As this "hollow" atom de-excites near the surface by Auger electron emission, electrons continue to flow from the metal to keep the ion neutralized. The emission of projectile K X-rays,[32] or of K-Auger electrons[33,34] as the inner-most vacancy is filled has been used as a "clock" to provide information about the formation and relaxation of such exotic atoms.

Both the K X-ray and K-Auger measurements indicate substantial occupation of the L shell when the 1s vacancy is filled, which cannot be explained by a complex step-wise decay of the hollow atom above the surface, since insufficient time is available for this to occur.[34] An explanation has been given recently by Meyer et al.[35] based on detailed measurements of ejected projectile K-Auger electrons from grazing interactions of N^{6+} ions with Au and Cu surfaces. Analysis of the measured electron energy spectra at different incidence angles and comparison with Monte-Carlo simulations permitted the identification of spectral features due to both above-surface and subsurface neutralization. The faster subsurface neutralization process dominates, populating the L shell directly, while the slower above-surface mechanism contributes only at grazing ion incidence angles, where sufficient time is available. The manifestation of hollow atoms may be more subtle than was originally thought!

SUMMARY AND OUTLOOK

The last two years have witnessed major advances in our understanding of collisional interactions of multiply charged ions which are important in high-temperature plasmas. This understanding has come from comparison of detailed experimental and theoretical results for simple interacting systems. Such few-electron systems are, for the most part, adequately described by theoretical approximations which have been implemented on existing computational facilities. The next challenge to our understanding will come as this arsenal of new experimental techniques and facilities is applied to progressively more complex interacting systems, where many-body interactions begin to dominate, and a myriad of surprises awaits us. This is sure to place severe demands on both the ingenuity of theorists and on supercomputer development.

REFERENCES

1. Y. Yongen and C. Lyneis, in *The Physics and Technology of Ion Sources*, edited by Ian G. Brown (Wiley, New York, 1989).

2. C.M. Lyneis and T.A. Antaya, Rev. Sci Instrum. 61, 221 (1990).

3. E.D. Donets in *Physics and Technology of Ions Sources*, edited by Ian G. Brown (Wiley, New York, 1989); Rev. Sci. Instrum. 61, 225 (1990).

4. R. Ali, C.P. Bhalla, C.L. Cocke and M. Stockli, Phys. Rev. Lett. 64, 633 (1990).

5. M.A. Levine, R.E Marrs, J.R. Henderson, D.A. Knapp and M.B. Schneider, Physica Scripta T22, 157 (1988).

6. D. Schneider, D. DeWitt, M.W. Clark, R. Schuch, C.L. Cocke, R. Schmieder, K.J. Reed, M.H. Chen, R.E. Marrs, M. Levine and R. Fortner, Phys. Rev. A 42, 3889 (1990).

7. P. Baumann, M. Blum, A. Friedrich, C. Geyer, M. Grieser, B. Holzer, E. Jaeschke, D. Krämer, C. Martin, K. Matl, R. Mayer, W. Ott, B. Povh, R. Repnow, M. Steck, E. Steffens and W. Arnold, Nucl. Imstrum. Meth. Phys. Res. A268, 531 (1988).

8. G. Kilgus, J. Berger, P. Blatt, M. Grieser, D. Habs, B. Hochadel, E. Jaeschke, D. Krämer, R. Neumann, G. Neureither, W. Ott, D. Schwalm, M. Steck, R. Stokstad, E. Szmola, A. Wolf, R. Schuch, A. Müller and M. Wagner, Phys. Rev. Lett. 64, 737 (1990).

9. L. H. Andersen, J. Bolko and P. Kvistgaard, Phys. Rev. Lett. 64, 729 (1990).

10. A. Müller, A. Frank, J. Haselbauer, G. Hoffman, J. Neumann, U. Pracht, E. Salzborn, S. Schennach, W. Spies, M. Stenke, O. Uwira, R. Völpel, M. Wagner, R. Becker, E. Jennewein, M. Kleinod, U. Pröbstel, R.A. Phaneuf, G.H. Dunn, E.M. Bernstein, N. Angert and P.H. Mokler, Proceedings of this Conference.

11. D.H. Crandall, R.A. Phaneuf, B.E. Hasselquist and D.C. Gregory, J. Phys. B. 12, L249 (1979).

12. D.H. Crandall, R.A. Phaneuf, D.C. Gregory, A.M. Howald, D.W. Mueller, T.J. Morgan, G.H. Dunn, D.C. Griffin and R.J.W. Henry, Phys. Rev. A 34, 1757 (1986).

13. K. Rinn, D.C. Gregory, L.J. Wang, R.A. Phaneuf and A. Müller, Phys. Rev. A 36, 595 (1987).

14. R.J.W. Henry and A.Z. Msezane, Phys. Rev. A 26, 2545 (1982).

15. M.S. Pindzola and D.C. Griffin, Phys. Rev. A 36, 2628 (1987).

16. G. Hoffman, A. Müller, K. Tinschert and E. Salzborn, Z. Phys. D 16, 113 (1990).

17. S.S. Tayal and R.J.W. Henry, Phys. Rev. A 42, 1831 (1990).

18. R. A. Phaneuf, "Experiments on Electron-Impact Excitation and Ionization of Ions, p.117-156 in *Atomic Processes in Electron-Ion and Ion-Ion Collisions*, ed. F. Brouillard, NATO Advanced Science Institutes Series B, Vol. 145 (Plenum, New York and London, 1986).

19. D.C. Gregory, L.J. Wang, K. Rinn and F.W. Meyer, Phys. Rev. A 35, 3526 (1987).

20. M.H. Chen, K.J. Reed, and D.L. Moores, Phys. Rev. Lett. 64, 1350 (1990).
21. E. Wåhlin, J.S. Thompson, G.H. Dunn, R.A. Phaneuf, D.C. Gregory and A.C.H. Smith, Phys. Rev. Lett. 66, 157 (1991).
22. B.A. Huber, Ch. Ristori, P.A. Hervieux and C. Guet, XVIIth International Conference on the Physics of Electronic and Atomic Collisions, Abstracts of Contributed Papers, Brisbane, Australia (1991), p.278.
23. D.C. Griffin and M.S. Pindzola, Phys. Rev. A 35, 2821 (1987).
24. M.B. Schneider, D.A, Knapp, M. Chen, J.H. Scofield, P. Beirsdorfer, C.L. Bennett, D.R. DeWitt, J.R. Henderson, M.A. Levine and R.E. Marrs, XVIIth International Conference on the Physics of Electronic and Atomic Collisions, Abstracts of Contributed Papers, Brisbane, Australia (1991), p. 304.
25. T. E. Cowan, C.L. Bennett, D.D. Dietrich, J.V. Bixler, C.J. Hailey, J.R. Henderson, D.A. Knapp, M.A. Levine, R.E. Marrs and M.B. Schneider, Phys. Rev. Lett. 66, 1150 (1991).
26. R. Hoeckstra, D. Ciric, F.J. de Heer and R. Morgenstern, Physica Scripta T28, 81 (1989); H.B. Gilbody, Physica Scripta T28, 49 (1989).
27. C.C. Havener, M.S. Huq, H.F. Krause, P.A. Schultz and R.A. Phaneuf, Phys. Rev. A 39, 1725 (1989).
28. C.C. Havener, F.W. Meyer and R.A. Phaneuf, XVIIth International Conference on the Physics of Electronic and Atomic Collisions, Invited Papers, Brisbane, Australia, 1991 (to be published).
29. N. Shimakura and M. Kimura, Phys. Rev. A 44, 1659 (1991).
30. H.D. Hagstrum, Phys. Rev. 96, 336 (1954); H.D. Hagstrum and G.E. Becker, Phys. Rev. B 8, 107 (1973).
31. U.A. Arifov, L.M. Kishinevskii, E.S. Mukhamadiev and E.S. Parilis, Zh. Tekh. Fiz. 43, 181 (1973) [Sov. Phys. Tech. Phys. 18, 118 (1973)].
32. J.P. Briand, L. de Billy, P. Charles, S. Essaba, P. Briand, R. Geller, J.P. Desclaux, S. Bliman and C. Ristori, Phys. Rev. Lett. 65, 159 (1990).
33. L. Folkerts and R. Morgenstern, Europhys. Lett. 13, 377 (1990).
34. P.A. Zeijlmans van Emmichoven, C.C. Havener, and F.W. Meyer, Phys. Rev. A 43, 1405 (1991).
35. F.W. Meyer, S.H. Overbury, C.C. Havener, P.A. Zeijlmans van Emmichoven and D.M. Zehner, Phys. Rev. Lett. 67, 723 (1991).

RECOMBINATION OF HIGHLY CHARGED IONS WITH FREE ELECTRONS

A. Müller, A. Frank, J. Haselbauer, G. Hofmann, J. Neumann, U. Pracht, E. Salzborn,
S. Schennach, W. Spies, M. Stenke, O. Uwira, R. Völpel, M. Wagner
Universität Giessen, W-6300 Giessen, Fed. Rep. Germany

R. Becker, E. Jennewein, M. Kleinod, U. Pröbstel
Universität Frankfurt, W-6000 Frankfurt, Fed. Rep. Germany

R. A. Phaneuf
Oak Ridge National Laboratory, Oak Ridge, TN 37830, USA

G. H. Dunn
Joint Institute for Laboratory Astrophysics, Boulder, CO 80309, USA

E. M. Bernstein
Western Michigan University, Kalamazoo, MI 49008, USA

N. Angert, P. H. Mokler
Gesellschaft für Schwerionenforschung (GSI), W-6100 Darmstadt, Fed. Rep. Germany

ABSTRACT

Recombination of highly charged ions and free electrons is studied in interacting-beams experiments. Beside direct recombination into bound states by radiative capture a variety of resonant recombination phenomena is observed. Resonant recombination produces a highly excited, usually short lived electron-ion compound which can stabilize by the emission of photons and/or electrons. Depending on this emission, the final charge state of the ion can be one less than the parent charge state, but it can also be higher and thus a net single or multiple ionization of the ion is observed after the initial recombination.

INTRODUCTION

Electron-ion collisions are fundamental processes which play an important role wherever ionized matter occurs. Data on electron-ion collision processes are needed for the modeling and understanding of astrophysical and laboratory plasmas like the solar corona, the discharge in a tokamak or laser produced plasmas. Among other electron-ion processes recombination is particularly interesting both from an applied point of view and as a testing ground for our understanding of basic atomic collision phenomena. The long-range Coulomb force allows for a complex variety of recombination phenomena the results of which can be found not only in the bound-state recombination of the electron and the ion but also in the channel of net ionization of the ion. Due to experimental difficulties inherent in the colliding-beams techniques direct measurements of electron-ion recombination cross sections and rates could be carried out only since a few years. Within a relatively short period of time, however, enormous experimental progress has been achieved due to technological advances in the design of electron targets, development of ion beam facilities, as well as in electronics for experimental control and data acquisition.

16 Highly Charged Ions with Free Electrons

A free electron and an ion A^{q+} can recombine by radiative recombination (RR)

$$e + A^{q+} \rightarrow A^{(q-1)+} + h\nu \tag{1}$$

where the excess energy is carried away by a photon in a direct process. After radiative recombination the captured electron can be in a highly excited (bound) state and hence further radiation will be emitted until the electron is in its ground level.

The recombination can also proceed in a resonant fashion

$$e + A^{q+} \rightarrow \left[A^{(q-1)+}\right]^{**} \tag{2}$$

where the excess energy released by the capture of the electron is used up by the excitation of a core electron within the ion. This so called resonant recombination or dielectronic capture can only occur if the kinetic energy of the projectile electron matches the difference $E_i - E_f$ of total binding energies of all electrons in the initial and final states of the ion. Inevitably dielectronic capture produces a multiply excited autoionizing state which can decay by the emission of photons

$$\left[A^{(q-1)+}\right]^{**} \rightarrow \left[A^{(q-1)+}\right] + nh\nu \tag{3}$$

and thus stabilize the reduced charge state of the ion which finally finds itself in the same electronic state as if it would have gone through radiative recombination. The whole two-step process of resonant recombination followed by stabilization via photoemission is termed dielectronic recombination.

Usually, for not too highly charged ions, electron emission from the intermediate resonant state

$$\left[A^{(q-1)+}\right]^{**} \rightarrow A^{q+} + e \tag{4}$$

is much more likely than photoemission so that the electron-ion compound disintegrates again into the original constituents. Beyond the two decay modes given by equations (3) and (4) there is also a probability for the emission of more than one electron after resonant recombination

$$\left[A^{(q-1)+}\right]^{**} \rightarrow A^{(q+m)+} + (m+1)e \tag{5}$$

and the net result is then an m-fold ionization of the ion A^{q+}. At sufficiently high electron densities three-body recombination becomes possible

$$e + e + A^{q+} \rightarrow A^{(q-1)+} + e \tag{6}$$

where one of the two electrons can carry away the excess energy released by the recombination. In this paper we concentrate on two-body collisions observed in experiments employing merged- or crossed-beams of electrons and ions.

Much of the activity on electron-ion recombination was devoted to Li-like ions. These ions provide a relatively simple electronic structure and correspondingly can be treated more easily by theory. Moreover, parent beams of such ions are usually free of metastable components which would make the interpretation of results difficult. These features make the lithium sequence attractive for experiments and therefore most of the examples of experimental results given in this paper are for Li-like ions. Exciting new results obtained by recombination experiments with many-electron ions both for the final bound-state and ionization channels are also discussed.

RADIATIVE AND DIELECTRONIC RECOMBINATION

For studying bound-state recombination of highly charged ions and free electrons we have set up a merged-beams experiment at the UNILAC accelerator of GSI in Darmstadt. The arrangement is schematically shown in Fig.1.

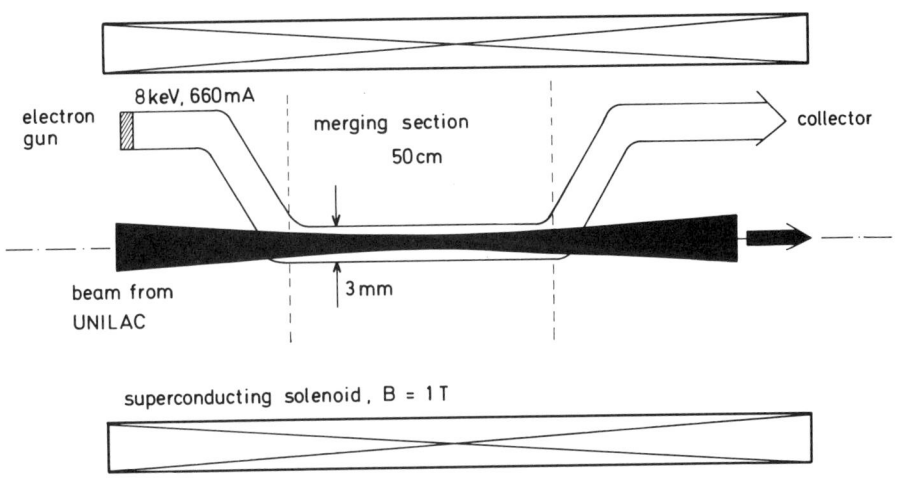

Fig.1: Schematic of our electron-ion merged beams setup.

Inside a strong magnetic field which is produced by a superconducting solenoid a dense electron beam is formed[1] serving as a target for an incident ion beam. The maximum electron density is $n_e=10^{10}$ cm^{-3} which is reached for an electron energy of 8 keV. Elliptical windings on top of the solenoid provide transverse magnetic field components which serve to merge and demerge the electron and the ion beams.

The electron target is incorporated in an ultra-high-vacuum beam line at the UNILAC accelerator which provides highly charged ions with energies up to 20 MeV/u. Charge-purified ion beams are axially injected into the electron beam. The recombined ions are magnetically separated from the parent ion beam and are detected by a position-sensitive multi-channelplate detector.

Fig.2 shows a spectrum of recombined Ar^{14+} ions formed in collisions of 5.9 MeV/u Ar^{15+} ions with free electrons whose laboratory energy was varied from 3300 eV to 3700 eV. Within this energy range the parallel velocities of both beams become equal so that at the matching point the center-of-mass energy is only determined by the velocity spreads in both beams. The cross section σ_{RR} for RR, which diverges when the energy approaches zero, leads to a finite experimental RR rate $r = <v\sigma_{RR}>$ given by the convolution of the total velocity spread with σ_{RR}. At about 3500 eV the parallel velocities of electrons and ions are matched and there a pronounced maximum in the total recombination rate is observed which is due to RR. We also see a number of well resolved DR resonances which can be associated with intermediate doubly excited states $(1s^2 2p_{1/2} n\ell)$ and $(1s^2 2p_{3/2} n\ell)$ of the Ar^{14+} ion with $n=10$ and 11. From the line shapes and the resolution of these measurements one can conclude that the maximum transverse velocity spread in the electron beam corresponds to about $kT_\perp = 0.2$ eV while the longitudinal velocity spread corresponds to $kT_\parallel=0.002$ eV.

18 Highly Charged Ions with Free Electrons

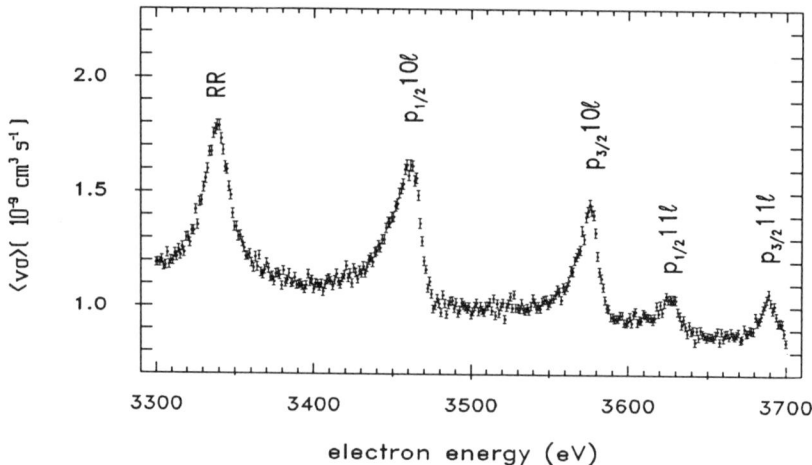

Fig.2: Experimental rates for recombination of 5.9 MeV/u Ar^{15+} ions and free electrons vs. electron laboratory energy. Background is not subtracted. The data are preliminary.

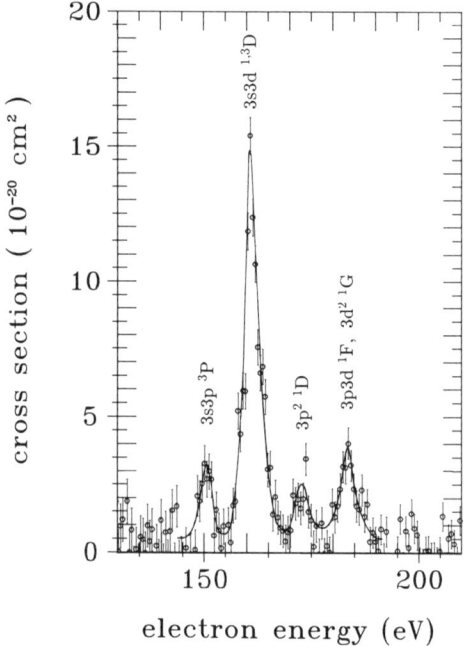

Fig.3: Experimental cross sections for dielectronic recombination of Ar^{15+} ions and free electrons via intermediate excited states $1s^2 3\ell 3\ell'$. The solid line was drawn to guide the eye. The data are preliminary.

The measurements for Ar^{15+} were extended up to the series limits of intermediate states with configurations $1s^2 2pn\ell$ and $1s^2 3\ell n\ell'$. The latter involve a $\Delta n=1$ transition $2s \to 3\ell$ of the active core electron which implies considerably smaller cross sections and rates than a $\Delta n=0$ transition. The resolution of the present experiment allows to separate individual terms in the $1s^2 3\ell 3\ell'$ configuration. Fig.3 shows a measurement of the $1s^2 3\ell 3\ell'$ resonance group populated by dielectronic recombination of Ar^{15+}. It is interesting to remark that few years ago such measurements of recombinations involving $\Delta n=1$ excitations were considered impossible with colliding-beams techniques[2].

New and unexpected phenomena were found in the recombination of highly charged many-electron ions with free electrons. When electron and ion velocities are matched in merged-beams experiments employing U^{28+} and Au^{25+} ions the measured recombination rates increase beyond any expectation. For U^{28+} Pindzola and Badnell[3] calculated $r = 1.8 \cdot 10^{-9}$ cm^{3}s^{-1} (for transverse and longitudinal energy spreads in the electron beam $kT_\parallel = 0.14$ eV and $kT_\perp = 6 \cdot 10^{-4}$ eV, respectively). The experiments, however, even with a slightly higher energy spread yield more than 10^{-7} cm^{3}s^{-1}.

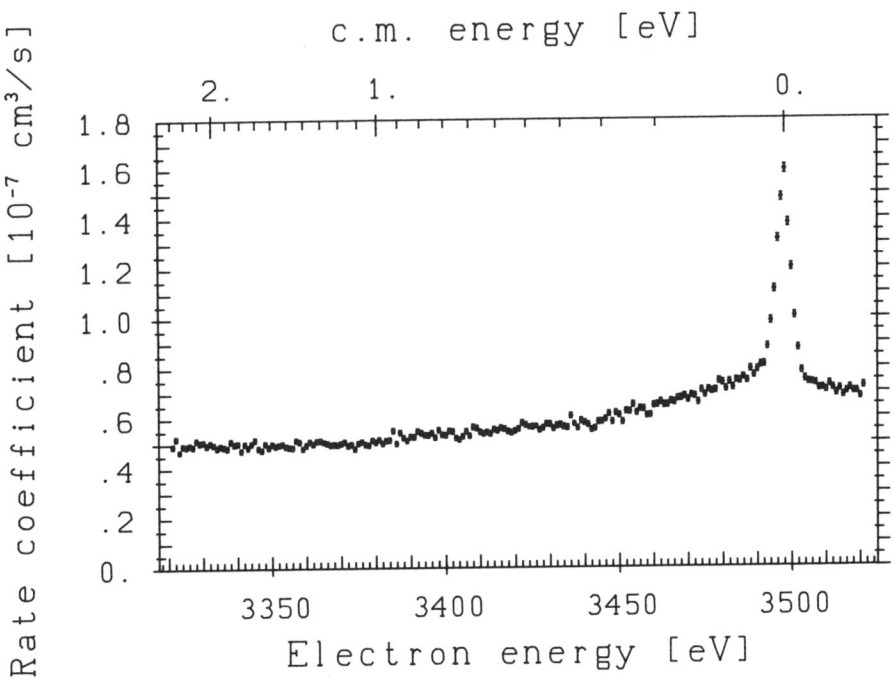

Fig.4: Rate for recombination of U^{28+} with free electrons as a function of the electron laboratory energy. Statistical uncertainties are indicated by vertical bars. Background is not subtracted.

Fig.4 shows a recombination-rate measurement for U^{28+} ions as a function of electron laboratory energy. At about 3500 eV the parallel velocities of electrons and ions are matched and there a pronounced maximum in the total recombination rate for e + $U^{28+} \to U^{27+}$ is observed. The energy dependence in Fig.4 suggests that there is a background contribution of $0.5 \cdot 10^{-7}$ cm^{3}s^{-1}. The peak rate after subtraction of this background amounts to more than $1 \cdot 10^{-7}$ cm^{3}s^{-1} and thus exceeds the expected rate by nearly a factor 100. One of the possible explanations is the presence of strong DR resonances on top of the RR peak, which would still be surprising because of the very high rates observed in the experiments. Further work is needed and will be carried out to test this hypothesis and to clarify the new observations.

Single-pass electron-ion merged-beams experiments like the present one are complementary to heavy-ion storage rings with an electron cooler where the ion beam repeatedly passes the electron beam. These rings provide outstanding possibilities to study a wide range of collision systems with high brightness and good energy resolution. The experimental program on electron-ion recombination at the Heidelberg storage ring TSR[4] has provided new data on a number of lighter ions (DR and RR were measured for ions in charge states up to 26+). In the near future we expect to carry out experiments with nearly completely stripped very heavy ions such as Bi^{80+} or U^{91+} employing the experimental storage ring ESR of GSI.

SINGLE AND MULTIPLE IONIZATION

As mentioned in the introduction the intermediate recombined states formed by dielectronic capture can be so highly excited that several electrons are emitted in a cascade of stabilizing transitions. The result is net single or even multiple ionization of the parent ion. Resonant recombination followed by autoionizations was first treated theoretically by LaGattuta and Hahn[5] who predicted strong contributions to total ionization cross sections of Fe^{15+} ions from resonances decaying by sequential emission of two electrons. Theory also predicted contributions to ionization from resonant recombination and subsequent double-Auger processes in which two electrons are emitted simultaneously from the electron-ion compound[6].

Only few years ago the predicted recombination resonances in net ionization could be experimentally observed. Beside the decay mechanisms where two electrons are emitted sequentially[7] or simultaneously[8] we found clear evidence of capture of a free electron with simultaneous excitation of two core electrons of the ion[9]. Moreover, we could demonstrate the occurence of recombination resonances in multiple ionization of ions by electron impact[7].

The difficulty to see recombination resonances in the electron-impact ionization of ions results from the presence of non-resonant "background" in the cross sections. This background is due to direct knock-off ionization of the ion by the projectile electron and due to the indirect process of excitation-autoionization where in a first step the ion is excited to an autoionizing state and then in a second step an electron is emitted. Since these non-resonant contributions usually dominate ionization cross sections by far one needs excellent statistics and energy resolution to find small resonance peaks on top of the smoothly varying cross section function. Attempts to overcome these problems usually fail because of the relative diluteness of beams of charged particles and because of the resulting low signal rates in a colliding beams experiment. Colliding-beams experiments, however, are the only key to low collision-energy spread and hence to good energy resolution in cross section measurements.

By using a fast energy-scanning technique as we have been doing in our bound-state-recombination experiments and by exploiting the potential of a high-intensity electron gun to produce high electron densities with energy spreads reduced by offsetting the negative electron space charge with slow positive ions we were able to measure ionization cross sections with relative uncertainties as low as 0.01% and energy resolutions $E/\Delta E$ of more than 200[7,9].

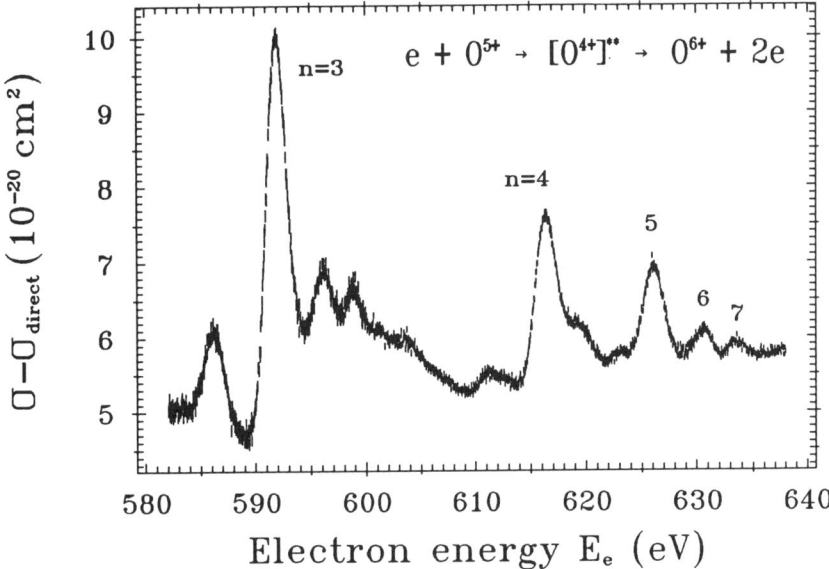

Fig.5: Experimental cross sections for indirect ionization of O^{5+} ions by electron impact. From the measured total cross section σ the contribution due to direct ionization σ_{direct} was subtracted in order to show the recombination resonances contributing in the particular energy range. The observed peaks are due to a Rydberg series $1s2s3\ell n\ell'$ of electron-ion compounds which decay by sequential emission of two electrons.

An example for a measurement of recombination resonances in net ionization of the parent ion is shown in Fig.5. The peaks correspond to a Rydberg series of $(e + O^{5+})$ compound states with configurations $1s2s3\ell n\ell'$ formed in ionizing collisions of electrons with O^{5+} ions. O^{6+} ions detected in the experiment can be produced for instance in a three-step process

$$e + O^{5+}(1s^22s) \to [O^{4+}(1s2s3\ell n\ell')]^{**} \to [O^{5+}(1s2s^2)]^{**} + e \to O^{6+}(1s^2) + 2e \quad (7)$$

The peaks shown in Fig.5 were obtained by accumulating about 10 million O^{6+} ions per energy with an energy spacing of about 0.04 eV. Hence, the relative uncertainty of each measured total ionization cross section is about 0.03 %. From the measured cross section function a straight line was subtracted extrapolating the direct-ionization contribution of the cross section into the energy range where indirect ionization processes occur. In the part of this range shown by the figure there is a smooth background of indirect ionization processes with a cross section of the order of $5 \cdot 10^{-20}$ cm^2. On top of this we see resonances resulting from an initial recombination of the free electron with the ion and subsequent two-electron emission. The dominating peak at 592 eV constitutes nearly 10 % of the total ionization cross section at the specific energy. It corresponds to one or several terms with a configuration $1s2s3\ell 3\ell'$. Members of the Rydberg series of intermediate resonant states $1s2s3\ell n\ell'$ are seen with principal quantum numbers up to

$n=7$. In these measurements the core excitation connected with the initial recombination process is related to $\Delta n=2$ transitions. Even recombination with $\Delta n=3$ transitions could be observed[10] in the ionization channel of electron-ion collisions. In the bound-state-recombination channel of electron-ion collisions no corresponding observations of such inter-shell transitions have been possible so far.

In a series of detailed measurements on heavy metal ions[7,11,12] narrow recombination resonances were found at identical energies in net single **and** in net multiple ionization of one given ion A^{q+}. Each set of these resonances produced at a given electron energy corresponds to one particular highly excited intermediate state of the ion $A^{(q-1)+}$ with its characteristic excitation energy. This state can decay by the emission of $(m+1)$ electrons and thus an ion $A^{(q+m)+}$ is produced with $m=1,2,...$. In the ionization of Ba^{2+} ions at about 800 eV electron energy, branching of resonant states into net m-fold ionization was observed for m up to 7, i.e. the intermediate resonant state in the Ba^{1+} ion emits up to 8 electrons.

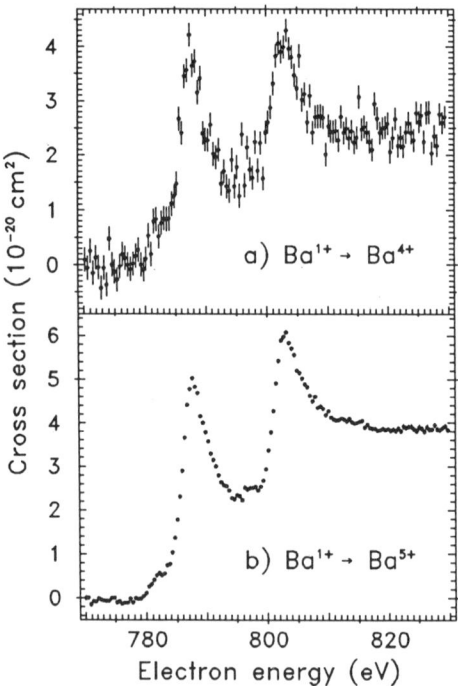

Fig.6: Comparison of energy-scan measurements[13] of net triple (a) and net quadruple (b) ionization of Ba^+ ions. Direct contributions to the cross section were represented by straight lines and subtracted from the measured total cross section in order to isolate the resonance features displayed here.

An example for such results is given in Fig.6 which shows features in cross sections of Ba^+ ions[13]. Here, the background arising from direct multiple ionization was determined from the total cross section measured below the resonance features. A straight line was fitted to these data and then subtracted from the total experimental cross section. What remains is a contribution involving the 3d subshell. There are two dominant peaks in both cross sections which are due to dielectronic capture plus subsequent emission of four or five electrons, respectively. Additional to the resonances there are contributions from 3d excitation with subsequent emission of three or four electrons, respectively, and probably also 3d ionization with subsequent emission of two or three electrons, respectively, as evidenced by an increased "background" level at energies above the resonances. All these processes can produce the net three-fold and net four-fold ionization observed in the experiments.

It is particularly interesting to note the different decay probabilities for the intermediate resonant 3d-excited states. Branching into net three-fold ionization yields a resonant cross section contribution of about $4 \cdot 10^{-20}$ cm^2 (for the peak at 787 eV) and branching into net four-fold ionization yields a cross section of about $5 \cdot 10^{-20}$ cm^2. This means that the branching ratio for the emission of five electrons from the resonant intermediate excited state exceeds that for the emission of four electrons by a factor of roughly 1.3 . When there is a vacancy in a deep lying shell the cascading Auger processes lead to a certain probability distribution of the final charge states. In the case of a 3d vacancy in an intermediate Ba atom the maximum of the probability distribution is around a charge state $q=5$. This can be understood in terms of an electron-evaporation model similar to the one used previously for describing transfer ionization processes in collisions of multiply charged ions and atoms[14].

Since the non-resonant contributions to the production of Ba^{q+} from Ba^+ ions are much higher for lower q it becomes increasingly difficult to find the low-q resonance contributions in experiments measuring the total cross section. This is also the reason why the statistical uncertainty of the data displayed in Fig.6 is higher for three-fold compared to four-fold ionization: the "background" from non-resonant multiple ionization is a factor 7 higher in $\sigma_{1,4}$ than in $\sigma_{1,5}$.

SUMMARY

Recombination of an ion A^{q+} with a free electron can lead not only to a final ion charge state $(q-1)$ resulting from photon-stabilized capture of the projectile electron into the target ion but also final charge states $(q+m)$ with $m = 0, 1, 2, 3, ...$ can be produced in processes where $(m+1)$ electrons are emitted from the intermediately formed electron-ion compound (see Fig.7). Thus, a spectroscopy of resonantly recombined states is possible in the channel of dielectronic recombination, in the channel of resonant elastic electron scattering and in the channels of single and multiple ionization. The origin of the peak features observed in the experiments is always a dielectronic capture (or resonant recombination) of the initially free electron into the parent ion, and it is only the decay mechanism which determines the branching into final charge states of the ion.

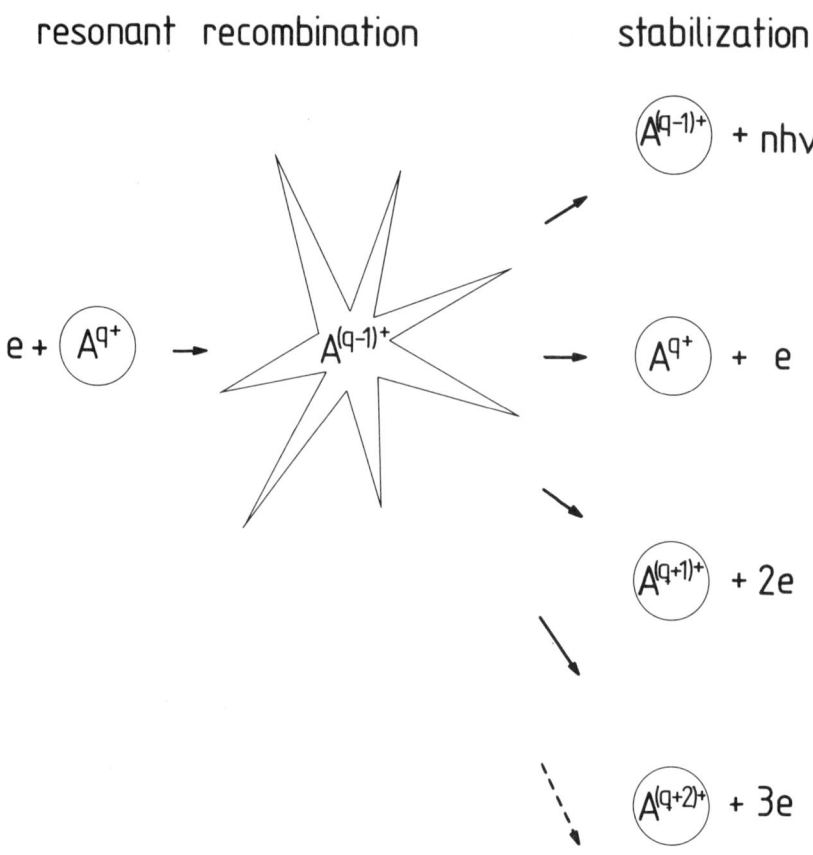

Fig.7: Scheme of resonant recombination of an electron with an ion and the subsequent stabilization transitions from the intermediate multiply excited compound state. The different decay modes lead to the following experimentally observable results: dielectronic recombination, resonant elastic (or inelastic) electron scattering, or resonances in single and multiple ionization.

ACKNOWLEDGEMENTS

We gratefully acknowledge the contributions made to this work by K. Tinschert, B. Weißbecker, N. Djurić, B. Jelenković, P. Spädtke, J. Klabunde, B. Wolf and staff members of GSI and of the universities in Giessen and Frankfurt.

The experiments on the bound-state-recombination channel reported here are supported by Bundesministerium für Forschung und Technologie (BMFT), and by Gesellschaft für Schwerionenforschung mbH (GSI), Darmstadt. Experiments on the ionization channel of resonant recombination are supported by Deutsche Forschungsgemeinschaft (DFG) and Max-Planck Institut für Plasmaphysik (MPP), Garching. NATO Collaborative Research Grant RG 86/0510 is gratefully acknowledged.

REFERENCES

1. M. Kleinod, R. Becker, E. Jennewein, U. Pröbstel, A. Müller, S. Schennach,
   J. Haselbauer, W. Spies, O. Uwira, M. Wagner, N. Angert, J. Klabunde,
   P. H. Mokler, P. Spädtke, B. Wolf
   Proceedings of the 19th INS International Symposium on *Cooler Rings and their Applications*, Tokyo, Japan, 5.-8.11.1990, editor T. Katayama, World Scientific Publishers, Singapore, 1991

2. D. C. Griffin, M. S. Pindzola, Phys. Rev. A **35**, 2821 (1987)

3. M.S. Pindzola and N.R. Badnell, private communication, 1990

4. A. Wolf, J. Berger, M. Bock, D. Habs, B. Hochadel, G. Kilgus, G. Neureither,
   U. Schramm, D. Schwalm, E. Szmola, A. Müller, M. Wagner, R. Schuch,
   in *Atomic Physics of Highly-Charged Ions*, eds. E. Salzborn, P. H. Mokler,
   A. Müller, (Springer, Heidelberg, 1991), in print as a Supplement Volume to
   Z. Phys. D Atoms, Molecules and Clusters

5. K. J. LaGattuta, Y. Hahn, Phys. Rev. A **24**, 2273 (1981)

6. R.J.W.Henry, A.Z.Msezane, Phys.Rev. A **26**, 2545 (1982)

7. A. Müller, K. Tinschert, G. Hofmann, E. Salzborn, G. H. Dunn,
   Phys. Rev. Lett. **61**, 70 (1988)

8. A. Müller, G. Hofmann, K. Tinschert, E. Salzborn,
   Phys. Rev. Lett. **61**, 70 (1988)

9. A. Müller, G. Hofmann, B. Weißbecker, M. Stenke, K. Tinschert, M. Wagner,
   E. Salzborn, Phys. Rev. Lett. **63**, 758 (1989)

10. G. Hofmann, A. Müller, K. Tinschert, E. Salzborn, Z. Phys. D **16**, 113 (1990)

11. A. Müller, K. Tinschert, G. Hofmann, E. Salzborn, G. H. Dunn, S. M. Younger,
    M. S. Pindzola, Phys. Rev. A **40**, 3584 (1989)

12. K. Tinschert, A. Müller, G. Hofmann, E. Salzborn, S. M. Younger,
    Phys. Rev. A **43**, 3522 (1991)

13. G. Hofmann, A. Müller, B. Weißbecker, M. Stenke, K. Tinschert, E. Salzborn,
    in *Atomic Physics of Highly-Charged Ions*, eds. E. Salzborn, P. H. Mokler,
    A. Müller, (Springer, Heidelberg, 1991), in print as a Supplement Volume to
    Z. Phys. D Atoms, Molecules and Clusters

14. A. Müller, W. Groh, E. Salzborn, Phys. Rev. Lett. **51**, 107 (1983)

DIELECTRONIC RECOMBINATION MEASUREMENTS OF HIGHLY-CHARGED HELIUMLIKE
AND NEONLIKE IONS USING AN ELECTRON BEAM ION TRAP*

Marilyn B. Schneider, David A. Knapp, P. Beiersdorfer, Mau H. Chen,
J.H. Scofield, C.L. Bennett, D.R. DeWitt, J.R. Henderson, Patricia
Lee+, Morton A. Levine++, R.E. Marrs, and D. Schneider
Lawrence Livermore National Laboratory, Livermore, CA 94550 USA
++Lawrence Berkeley Laboratory, Berkeley, CA 94720 USA

*Work performed under the auspices of the U.S. Deptartment of Energy by the
Lawrence Livermore National Laboratory under contract number W-7405-ENG-48.
+now at Georgia Institute of Technology, Atlanta, GA 30332 USA

ABSTRACT

The electron beam ion trap (EBIT) at LLNL is a unique device designed to measure the interactions of electrons with highly-charged ions. We describe three methods used at EBIT to directly measure the dielectronic recombination (DR) process : (1) The intensity of the stabilizing X rays is measured as a function of electron beam energy; (2) The ions remaining in a particular ionization state are counted after the electron beam has been held at a fixed electron energy for a fixed time; and (3) High-resolution spectroscopy is used to resolve individual DR satellite lines. In our discussions, we concentrate on the KLL resonances of the heliumlike target ions (V^{21+} to Ba^{54+}), and the LMM resonances of the neonlike target ions (Xe^{44+} to Th^{80+}).

INTRODUCTION

Dielectronic recombination (DR) is an important process in fusion plasmas, solar flares, and astrophysical systems because it strongly influences the ionization balance and emitted x-ray spectra. Measurements of DR in highly-charged ions are also of fundamental interest because they can test multi-electron atomic physics models in the high-Z, highly relativisitic limit. The EBIT machine[1-3] at LLNL is an ideal device in which to measure the DR resonances of highly-charged ions[4-9]. In this paper we report several complementary techniques used to measure DR of heliumlike and neonlike target ions.

DR is the resonant capture of electrons by ions. An intermediate state is formed when the energy gained by capturing an electron from the continuum resonantly excites one of the bound electrons. DR occurs when the ion, now with one less charge, decays by emitting a photon.

The DR resonances can be seen with x-ray or particle techniques by measuring either the intensity of the DR X rays or the ion abundance as a function of the interaction energy between the ion and the electron. We have previously used x-ray techniques to measure the KLL DR resonances of heliumlike Ni^{26+} [4-5] and Mo^{40+} [5] target ions and the LMM resonances of neonlike Au^{69+} [6] target ions. We summarize these results and also report new results for heliumlike Ba^{54+} and neonlike Xe^{44+} and Th^{80+} target ions. The cross sections determined with the x-ray technique are normalized to those of radiative recombination (RR), the nonresonant recombination of electrons with ions. A particle technique has previously been used to measure the abundance ratio, Ar^{16+}/Ar^{15+}, of ions extracted from an electron beam ion source.[10] In this case, the DR cross sections were normalized to those of electron impact ionization. We also measure ion abundances by using the EBIT machine as an ion source.[3] Using ion-extraction techniques, we have measured DR of hydrogenlike Ar^{17+} target ions[8] and, more recently, neonlike Xe^{44+} target ions[9]. We summarize the latter experiment here. The extraction data is not normalized to obtain a cross section, but the relative strengths of the resonances are measured with the highest electron energy resolution yet from EBIT. We note that a process similar to DR has also been measured in resonant transfer and excitation (RTE) experiments where ions move through a gas target and interact with the target's electrons. The RTE equivalent of the KLL resonances of heliumlike U^{90+} onto H_2 was observed[11].

We confine our attention in this paper to the KLL resonances of heliumlike target ions and the LMM resonances of neonlike target ions. These resonances are the strongest for the highly-charged ions which we consider. As an example, the LMM resonances in neonlike gold target ions are expected to have about 70% of the total DR resonance strength.

We note that the LMM resonances in neonlike ions differ from the KLL resonances in heliumlike ions in several ways: (a) the energy threshold for ionization to a neonlike state is 2-3 times the energy of the LMM resonances, whereas the threshold for ionization of a heliumlike state is below the energy of the KLL resonances. This means that extraction experiments on neonlike target ions cannot be normalized to an ionization cross section. (b) the LMM resonance strengths are at least an order of magnitude larger than the KLL resonance strengths. This means that the LMM experiments must be performed on a shorter timescale, because the number of neonlike ions is quickly depleted. (c) the LMM resonances of neonlike target ions have 237 intermediate states whereas the KLL resonances of heliumlike target ions have only 16.

THE ELECTRON BEAM ION TRAP

The electron beam ion trap (EBIT) is a device specifically designed to study the X rays produced by very-highly-charged ions when they interact with free electrons[1-2]. The ions are available for study because they are electostatically trapped for seconds to hours in a narrow cylindrical region defined by biased electrodes in the axial direction and by a high-current-density electron beam in the radial direction. The energy of the electron beam can be varied reliably from 2 keV to about 40 keV with a typical energy spread of 50 eV FWHM. X rays from these interactions are observed at 90° to the electron beam direction. Recently, EBIT was upgraded to allow the ions to be extracted from the trap and magnetically analyzed by charge-to-mass ratio.[3]

DR OF NEONLIKE TARGET IONS USING THE "EVENT-MODE" X-RAY TECHNIQUE

Typically, the DR resonances are measured using an "event-mode" data acquisition technique and a solid-state detector. A schematic diagram of the EBIT voltages for an experiment on neonlike target ions is illustrated in Fig. 1. First the ions are dumped out of the trap by turning off the electron beam and inverting the axial

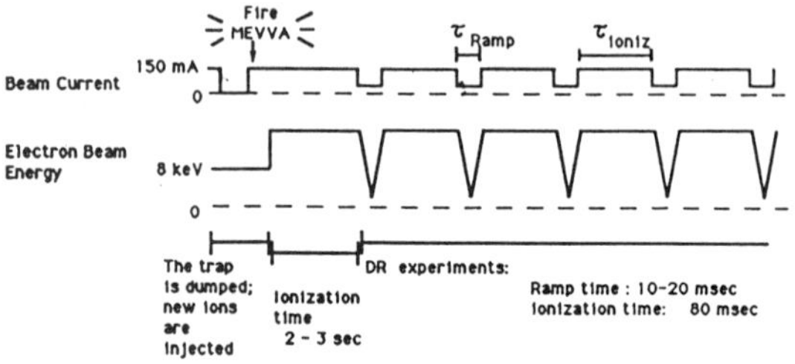

28 Dielectronic Recombination Measurements

Fig. 1 Typical timing pattern for the electron beam current and the electron beam energy in an event-mode data acquisition experiment in EBIT. Parameters shown are those used for the neonlike Au experiment.

potential. Then, the beam is turned on, the axial trap is restored, and low-charged ions (q=+1 to +4) are injected into EBIT. These ions are successively ionized by the electron beam under conditions which maximize the ratio of neonlike ions to sodiumlike ions. This requires a high beam current and an electron beam energy that is just below the threshold for removing the 10th electron. Once the ionization balance is established, the DR experiment is performed. Dielectronic recombination is a strong recombining process, so setting the electron energy to the resonant energy would soon destroy the ionization balance. The resonances are probed by reducing the beam current and linearly ramping the energy of the electron beam down from the ionization energy to an energy below that of the dielectronic recombination resonances, and back up. The ionization conditions are then restored to maintain the ionization balance. The cycle is repeated about 200 times before the trap is again dumped. Note that the time spent on resonance (typically 17 uS for a 3keV/mS ramp) is too short to perturb the ionization balance at the reduced beam currents used.

X rays are observed during ramping with a solid-state Ge or Si(Li) detector. When an X ray is detected, its photon energy, the voltage used to accelerate the electron beam, and the time are recorded. If the voltage ramp is linear, the data is automatically normalized because the same amount of time is spent at each voltage.

The data from a typical run can be displayed as a scatter plot, as shown in Fig. 2 for neonlike gold. X-ray spectra are horizontal slices of this plot. Excitation functions (x-ray intensity versus electron energy) are generated by projecting selected events onto the electron beam energy axis. The DR excitation function of the data which is analyzed is the sum, for each electron energy, of all RR X rays to n=3 plus all n=3->2 DR X rays. Because the intermediate state of the ion has two electrons in the n=3 shell but only one hole in the n=2 core, there is only one n=3->2 X ray emitted for each recombination. The excitation functions are shown in Fig. 3a,b,c for the xenon, gold, and thorium, systems, respectively. As the target ion becomes more highly charged, individual DR resonances become easier to resolve because the spacing between the electron energy levels increase. However, the ionization balance in EBIT worsens, as evidenced by an increase in the number of peaks in the thorium data compared to the gold data. This is because the ionization cross section of sodiumlike ions scale like $1/Z^4$ while the recombination cross sections of neonlike ions are either approximately constant (radiative recombination) or scale like Z (charge exchange) so that the ratio of the number of neonlike ions to the number of sodiumlike ions in EBIT, scales, at best, like $1/Z^4$.

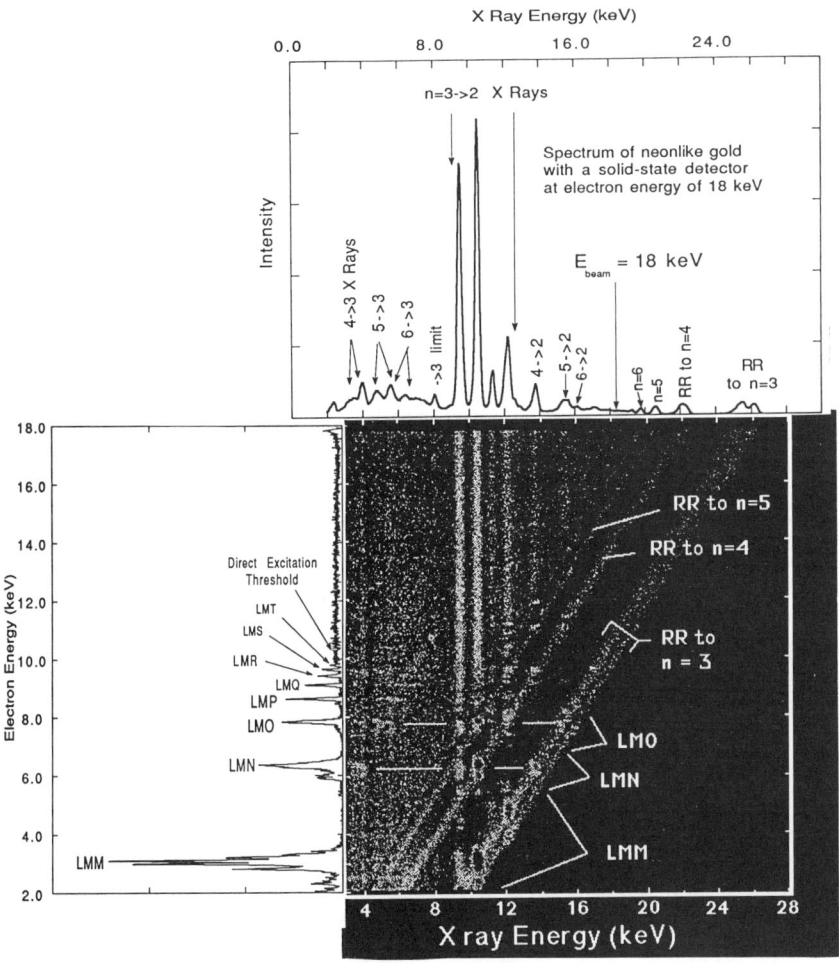

Fig. 2 The data from a DR experiment on neonlike gold target ions displayed in a scatter plot (middle figure) which maps x-ray intensity on a greyscale versus x-ray energy (x-axis) and electron beam energy (y-axis). X-ray excitation lines appear as vertical bands; radiative recombination (RR) X rays appear in diagonal bands. The high intensity peaks at the intersection of the excitation lines and RR bands are the DR resonances. A horizontal cut through the scatter plot gives an x-ray spectrum at a fixed electron energy. This is illustrated in the top figure for the ionization energy of 18 keV. The n=3->2 x-ray excitation lines are identified. A vertical cut along a fixed x-ray line gives a DR excitation function. The figure on the left is a vertical cut through the brightest n=3->2 X-ray line. The DR resonances are labeled. The strongest resonances are the LMM resonances.

Note that a diagonal cut along an RR band gives a complementary DR excitation function.

A multiconfiguration Dirac-Fock (MCDF) calculation[12-13] of the resonance strengths and positions are used to fit each excitation function for an energy offset, an electron beam energy resolution, and an amplitude of each target ionization state for up to four target ionization states (neonlike to aluminumlike). Accurately calculated radiative recombination cross sections[14] are used for the background term. For the calculation of the DR resonance strengths, the target ions were assumed to be initially in their ground state because the electron density was so low ($< 3 \times 10^{12} cm^{-3}$) that the time between an electron-ion collision was longer than the lifetime of any metastable state of the ion. The atomic energy levels and bound-state wave functions were calculated using the MCDF model in the extended average-level scheme.[12] The effects of quantum-electrodynamic corrections, the finite nuclear size, the transverse Breit interaction, and relaxation were included in the calculations of the transition energies. The relativistic Auger and radiative rates for each intermediate state were calculated using first-order perturbation theory. Because the DR resonances have much narrower widths than the beam resolution, they

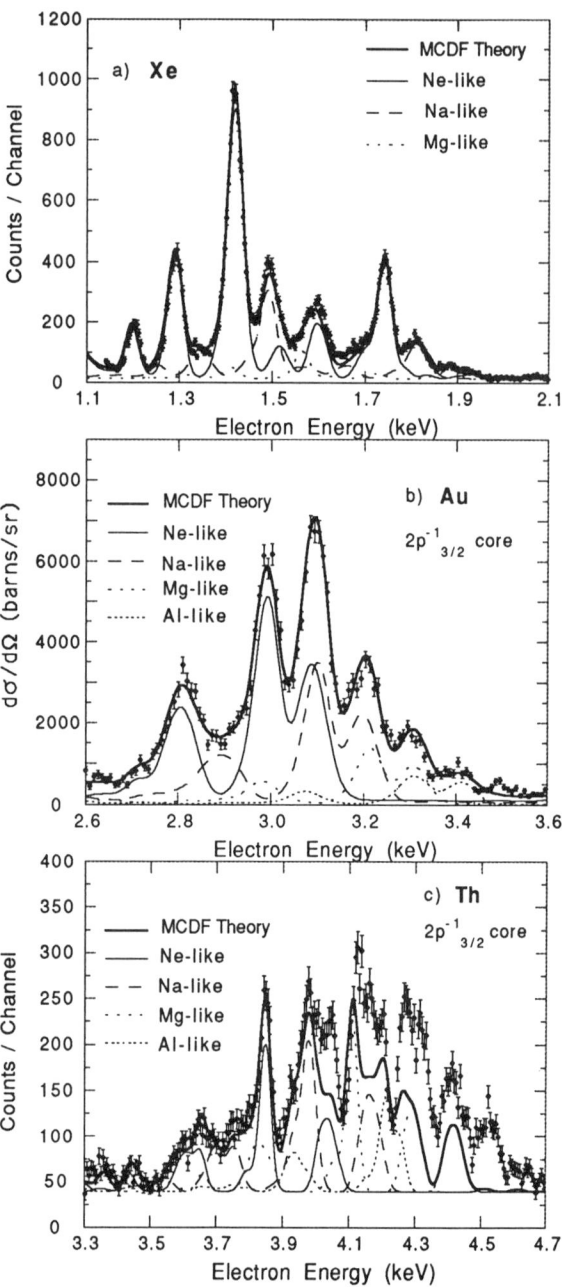

Fig. 3 The excitation functions for the strongest portion of the LMM resonances for neonlike (a) Xe^{44+}, (b) Au^{69+}, and (c) Th^{80+} target ions, respectively.

were treated as δ functions normalized to the energy-averaged cross sections as defined by Lagattuta and Hahn[15]. The cross sections for emission of X rays at 90° to the electron beam were expressed as resonance strengths averaged in energy bins of 1 eV. A more complete presentation of the techniques used is given in Ref. 13.

The theoretical fit to the Xe data required three target ionization states, the Au data needed four states, but the Th data requires at least six states (which were not yet available). The results are shown in Fig. 3, along with the contribution of each target ionization state. The shapes of the resonances for the Xe and Au data are in very good agreement with theory. The part of the Th data involving the neonlike target state is also in good agreement with theory. The Gaussian FWHM of the electron energy was 30 eV for the xenon data, 60 eV for the gold data, and 40 eV for the Th data. The Au data has the poorest resolution because of noise on the high-voltage power supply used to accelerate the electron beam in that experiment. The Th data has slightly worse resolution than the Xe data because it was obtained at a beam current of 50 mA, and the Xe data was obtained with a 30 mA current.

To obtain an absolute cross section, the overlap of the ions with the electron beam must be known. Because this is not possible, the strength of the dielectronic recombination resonances were normalized to that of radiative recombination. The normalization factor, in counts per cross section, is obtained by dividing the number of RR X rays at each electron energy by the detector efficiency times the sum of the fraction of each ionization state (obtained from the theoretical fit to the DR excitation function) times its calculated radiative recombination cross section. The RR cross sections for X rays emitted at 90° to the electron beam for each ionization state were calculated using a relativistic distorted-wave code.[14] Care was taken to use only those regions of the RR bands to n=3,4,5 which were uncontaminated by other processes.

About 20 out of 50 LMM DR resonances onto the neonlike ion, 40 out of 115 onto the sodiumlike ion, 20 out of 60 onto the magnesiumlike ion, and 40 out of 160 onto the aluminumlike ion make significant contributions to the DR cross sections for the Au and Th data shown in Fig. 3. The dominant contribution is from high angular momentum intermediate states with a $2p^{-1}{}_{3/2}$ core and at least one DR electron in the $3d_{5/2}$ subshell.

The final measured total resonance strength for DR in neonlike gold between 2.6 and 3.6 keV is $1.0 \pm 0.15 \times 10^{-17}$ cm^2-eV; the theoretical value is 0.95×10^{-17} cm^2-eV. The principal source of error comes from the uncertainty in the normalization factor.

We have also carefully examined the LMM resonances in neonlike gold with intermediate states with $2p^{-1}{}_{1/2}$ and $2s^{-1}{}_{1/2}$ core configurations. Figure 4 shows the data in this region compared to the MCDF theory with the ionization balance, beam width, and normalization factor from the data in Fig. 3b. A discrepancy was found for the peak at 4.55 keV. The MCDF theory predicts this peak to be composed of two neonlike DR resonances: 80% from an intermediate state composed mainly of the $(2p^{-1}{}_{1/2}3p_{3/2}3d_{3/2})_{J=5/2}$ state and 20% from an intermediate state composed mainly of the $(2s^{-1}{}_{1/2}3s_{1/2}3d_{5/2})_{J=5/2}$ state. It is likely that the theory

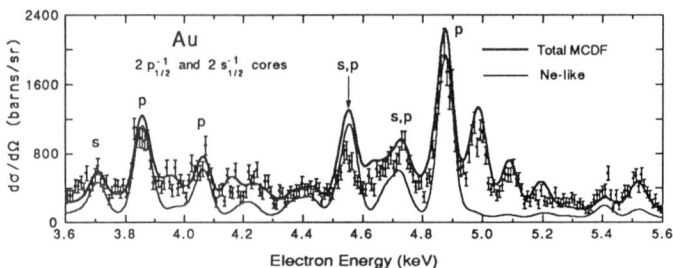

Fig. 4 DR excitation function data of neonlike Au^{69+} target ions for the $2s^{-1}_{1/2}$ and $2p^{-1}_{1/2}$ core regions. The theory uses the parameters from Fig. 3b. The major neonlike peaks are identified by s or p for the different cores. The arrow identifies the peak with biggest discrepancy with the MCDF theory.

overestimates the mixing between these two states, making the resonance of the mainly $2s^{-1}_{1/2}$ state too large.

EXTRACTION EXPERIMENTS

We have also measured the DR resonances of neonlike Xe^{44+} using EBIT as an ion source.[9] In these experiments, an ionization balance is first established, and then the DR resonances are probed by lowering the beam current (typically to 5 mA) and the beam energy (to a probe energy) for a given probe time.

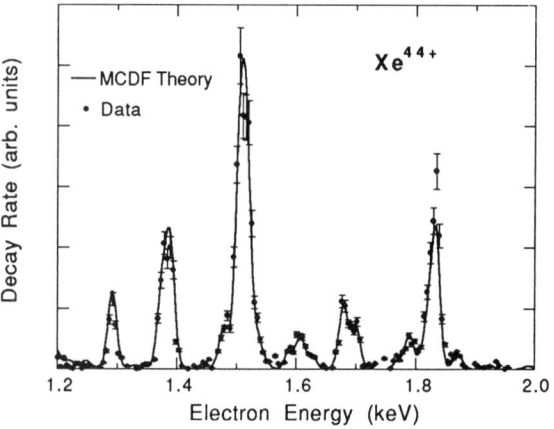

Fig. 5. Decay rate of the number of Xe^{44+} ions extracted versus probe electron energy. The smooth curve is the MCDF theory for the DR resonance strengths convoluted with an electron energy resolution of 16 eV FWHM.

The beam current is then turned off and the beam energy increased to the fixed extraction energy. The ions are extracted by pulsing the axial trap to invert it. Those ions with a preselected charge are magnetically steered into a photomultiplier tube and counted. Two to five dumps per probe condition are usually performed, and then the probe energy or the probe time is changed. A plot of the number of neonlike Xe^{44+} ions versus probe energy shows a large decrease in the number of Xe^{44+} at its DR resonances. At each probe energy, the number of ions decreases exponentially with time because the only process occuring is recombination, so the experiment is repeated for several different probe times. For each probe energy, the data at different probe times is used to find the exponential decay rate. Fig. 5 shows the decay rate versus electron energy. This decay rate is proportional to the sum of the different recombination processes, dielectronic recombination, radiative recombination, and charge exchange with background gases. The latter two processes vary smoothly with energy, so the theoretical DR cross sections are used to fit the data in Fig. 5 for a smooth background, an electron energy resolution, and a scaling factor. The electron energy resolution for the data in Fig. 5 is 16 eV FWHM, as compared to the 30-eV FWHM for xenon data in Fig. 3a. The improved resolution is due to the low beam current and low axial trapping voltage (~10 V) used. While these two factors reduce the number of ions in the beam compared to an x-ray technique, there is still plenty of signal in the extracted ions. Under these conditions, the variance of the electron energy in the overlap of the ion cloud and electron beam is smaller, improving the resolution.

NEW RESULTS ON HELIUMLIKE SYSTEMS

We have also used the "event-mode" data acquisition technique to measure the KLL resonances of heliumlike Ni^{28+}, Mo^{44+} and Ba^{54+} target ions. Fig. 6 shows the result. Again, the agreement with theory is excellent. The Ni and Mo data have been normalized. The electron energy resolution is 50-60 eV FWHM for this data. Some structure is resolved in the Mo and Ba data.

Recently, we obtained high resolution x-ray spectra of the DR resonances in heliumlike V^{21+}.[7] We used a Bragg crystal spectrometer in the von Hamos geometry[16] to record spectra at fixed electron energies, for values of the electron energy at and near the DR resonances. Since the KLL resonances are ABOVE the ionization threshold of Li-like to He-like ions in heliumlike systems, a steady-state ionization balance is reached at each electron energy. (Especially at electron energies near the DR resonances, the

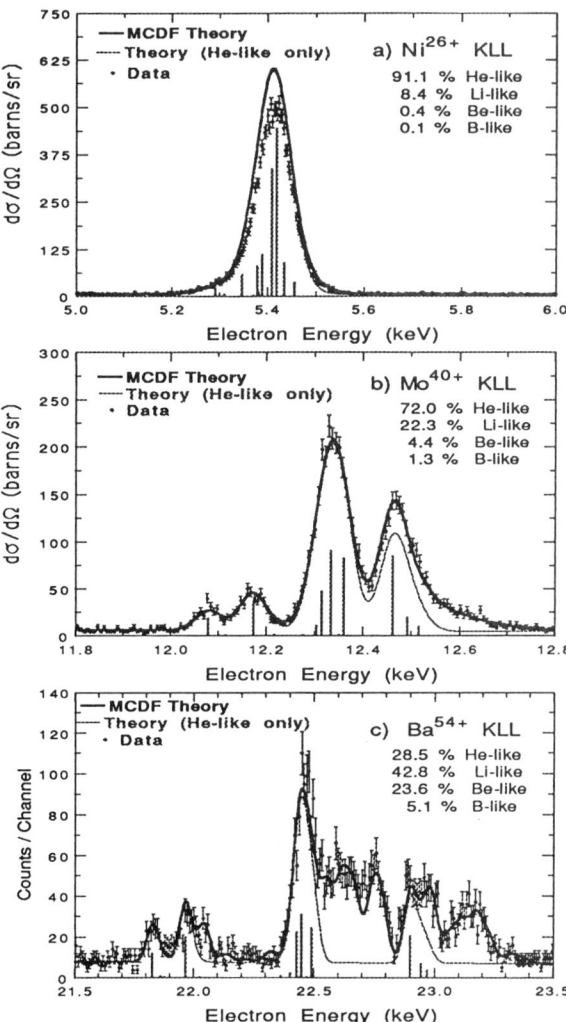

Fig. 6. Comparison of theory and experiment for the KLL DR resonances of heliumlike target ions using the event-mode x-ray technique. The vertical lines represent the theoretical resonance strengths of the heliumlike target ion. The curves are the theoretical resonance strengths folded with the experimental electron energy resolution. Theory curves are shown both for heliumlike target ions only, and for the actual ionization balance present in the trap. The measured amplitude is a) Ni^{26+} 85 ± 7 %; b) Mo^{40+} : 102 ± 6 % of the theoretical value.

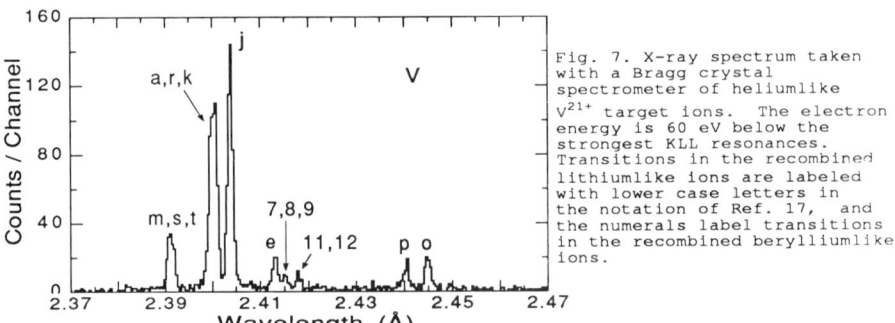

Fig. 7. X-ray spectrum taken with a Bragg crystal spectrometer of heliumlike V^{21+} target ions. The electron energy is 60 eV below the strongest KLL resonances. Transitions in the recombined lithiumlike ions are labeled with lower case letters in the notation of Ref. 17, and the numerals label transitions in the recombined berylliumlike ions.

ionization balance is poor). Figure 7 shows a typical x-ray spectrum of V at an electron energy near the KLL resonances. The lithiumlike satellite line positions were measured with an average accuracy of 0.2 mÅ, which is a resolution ($\Delta E/E$ or $\Delta\lambda/\lambda$) of 8×10^{-4}. The lithiumlike satellite lines e, o, and p were seen for the first time in such highly-charged ions.

We have added the von Hamos spectrometer to the event-mode data acquisiton system. We have used this to study the DR resonances of heliumlike Cr^{22+} target ions. Figure 8 shows the scatter plot and an x-ray spectrum at the KLL resonances. This technique preserves the original ionization balance.

SUMMARY

In summary, we have discussed several techniques used to measure DR resonances using an electron beam ion trap. The event-mode x-ray technique with a solid state detector gives an electron energy resolution of 50 eV FWHM, but poor x-ray energy resolution. However, the DR resonance strength relative to that of radiative recombination can be determined because the data acquisition process is self-normalizing. The event-mode coupled to a von Hamos spectrometer gives the same electron energy resolution but the resonances are resolved to 8×10^{-4} in x-ray energy. These x-ray experiments benefit from high beam currents to improve the data rate. The extraction technique, which uses EBIT as an ion source, allows very high resolution in electron energy (16 eV FWHM) because it can use a low beam current and low trapping voltage. Its disadvantage is the inability to normalize the resonance strengths.

REFERENCES

1. M.A. Levine et al., Phys. Scr. **T22**, 157 (1988).
2. M.A. Levine et al., Nucl. Instr. and Meth. **B43**, 431 (1989).
3. D. Schneider et al., Phys. Rev. A **42**, 3889 (1990).
4. D.A. Knapp et al., Phys. Rev. Lett. **62**, 2104 (1989)
5. D.A. Knapp, proceedings of Vth Inter. Conf. on the Physics of Highly-Charged Ions, Giessen, FRG, September 10-14, 1990, Z. Physik. D (in press).
6. M.B. Schneider et al., submitted to Phys.Rev.Lett. 6/91.
7. P. Beiersdorfer et al., Phys. Rev. A. **44**, 396 (1991).
8. D.R. DeWitt et al., submitted to Phys.Rev. A 7/91.
9. D.R. DeWitt et al., submitted to Phys. Rev. Lett. 8/91.

10. R. Ali et al., Phys. Rev. Lett. **64**, 633 (1990); Phys. Rev. A. **44**, 223 (1991).
11. W.G. Graham et al., Phys. Rev. Lett. **65**, 2773 (1990)
12. I.P. Grant, et al., Comput. Phys. Commun. **21**, 207 (1980).
13. M.H. Chen, Phys. Rev. A **31**, 1449 (1985); **33** 994 (1986).
14. J.M. Scofield, Phys. Rev. A **40** 3054 (1989); E.B. Saloman et al., At. Data Nucl. Data Tables **38**, 1 (1988)
15. K. LaGattuta, Y. Hahn, Phys.Rev. A **24**, 2273 (1981); J.Phys.B **15**, 2101 (1982).
16. P. Beiersdorfer et al., Rev. Sci. Instrum. **61**, 2338 (1990).
17. A.H. Gabriel, Mon. Not. R. Astron. Soc. **160**, 99 (1972).

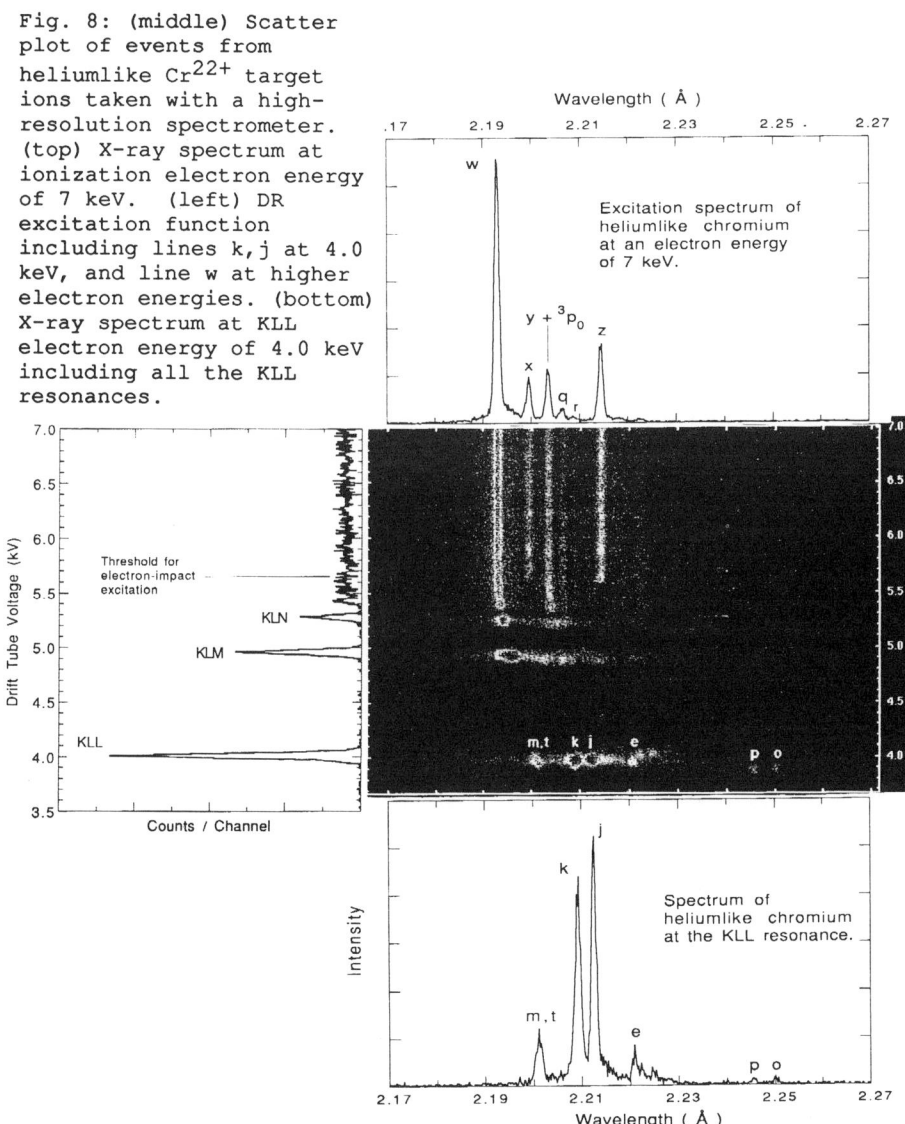

Fig. 8: (middle) Scatter plot of events from heliumlike Cr^{22+} target ions taken with a high-resolution spectrometer. (top) X-ray spectrum at ionization electron energy of 7 keV. (left) DR excitation function including lines k,j at 4.0 keV, and line w at higher electron energies. (bottom) X-ray spectrum at KLL electron energy of 4.0 keV including all the KLL resonances.

DENSE PLASMAS

A REVIEW OF SPECTRAL LINE BROADENING RELEVANT TO HOT DENSE PLASMAS

R. W. Lee

Lawrence Livermore National Laboratory L-23, POB 808 Livermore, CA 94550

ABSTRACT

A review of the recent work performed in the study of spectral line broadening in dense plasmas is presented. The work reviewed will cover recent advances in the theoretical development of line broadening and touch only briefly on the experimental aspects of the problem. Efforts to develop line shape calculations appropriate for spectral synthesis, which can be used in the study of opacities and non-LTE radiative transfer will also be discussed. A brief summary of the currently available resources for line shapes will be presented. A bibliography of the papers used in the review is provided.

I. INTRODUCTION

The study of spectral line broadening in high density plasmas forms an important component of the plasma spectroscopy of both laboratory and astrophysical plasmas. For the purposes of this brief review the working definition of a high density plasma will be a plasma in which the line profiles are dominated by the Stark broadening due to charged perturbers interacting with ionic emitters. This excludes plasmas where the Doppler or natural lifetime broadening dominates; but, includes a wide regime of charged emitter species, Z, electron densities, N_e, and temperatures, T_e.

Over the past few years experimental high density plasma studies have been dominated by those plasmas created for inertial confinement fusion, x-ray lasers studies, and direct laser-matter interaction studies. This defines one direction for development of both theory and experiment. A second direction for spectral line broadening of ionized emitters arises from the need for spectral synthesis relevant to a new generation of non-LTE kinetics studies, detailed radiative transfer simulations and multi-frequency opacity calculations The latter development is prompted mainly by astrophysical applications but has direct relevance to laboratory plasmas. This new direction requires the line broadening calculations to treat arbitrarily complex atomic structures while keeping calculation time as low as possible.

In choosing the topics for review a decision was made to exclude the radiative transfer effects on line profiles. Although this area of study is rapidly evolving, with new more robust techniques becoming available, intrinsic line shapes are, to a large degree, separable from the radiative transfer process. In this sense the line shapes can be seen as data to be provided to the radiative transfer scheme and not intrinsically a part of the scheme. The one obvious problem in this separation would arise in the treatment of detailed frequency redistribution. However, leaving this area of radiative transfer aside, the separate study of spectral line profiles formed by Stark broadening is an ample field of research in itself.

In the following section, a very brief and schematic derivation of the standard line shape theory will be given. This theoretical starting point can be used to discuss some recent advances made in the study of spectral line broadening. A brief outline of the advances is given. In the section III comments on recent scattering calculations are presented. Although scattering calculations provide line width and shift estimates, *not* line shapes, they are central to the work performed over the past few decades. A discussion is then presented of the stochastic, or microfield, methods which are the one theoretical development that is considered both novel and may provide an improved treatment of line shapes in dense plasmas.

The recent interest in spectral synthesis for non-LTE plasma studies arising from astrophysics and x-ray lasers simulations, along with the desire for low and mid-Z opacities, has given rise to a new development in spectral line broadening studies. Work in this area emphasizes the rapid calculation of accurate profiles for arbitrarily complex emitters and has allowed line profile prediction for species that were heretofore beyond the scope of line shape calculation. This work is discussed and examples are presented.

A comment on the experimental work in line shapes is then given. Here the difficulty in producing plasmas with the appropriate supporting diagnostic information limits the number of experiments. The fact that line shapes, or in most cases, line widths can provide a diagnostic where other methods fail is the other side of the fact that producing interesting and well diagnosed plasmas for the study of high density line shapes is hard, and possibility unrewarding, work.

The future prospects for the study of high density line shapes is presented in section IV. Here the one clear prospect, that is the future of the stochastic theories, is discussed. Comments on two of the outstanding, and outstandingly difficult, problems faced in the study of spectral line broadening, line wings and consistent non-LTE effects are pointed out. Finally, a list of resources for line shapes and a brief bibliography are presented.

II. THEORETICAL STARTING POINT

The starting point for the formulation of spectral line broadening in plasmas will be, what is called here, the standard theory. In this model it is assumed that the ions can be treated as a slow quasi-static set of perturbers, while the electrons can be treated as rapid impacts. This solution of the line broadening problem was first presented in 1958.[1] Defining the lifetime of a transition as $\tau_{1/2}$ and using the Fourier theorem to relate the half-width of the line shape, $\omega_{1/2}$, to the lifetime as

$$\omega_{1/2} \tau_{1/2} \sim 1 \qquad (1)$$

the justification for this separation of ion and electron effects can be made. That is, since, *on average*

$$\tau_{1/2} \gg \tau_{\text{e-collision}} \quad \text{and} \quad \tau_{1/2} \ll \tau_{\text{ion collision}} \qquad (2)$$

where $\tau_{\text{collision}}$ is the collision duration, one can roughly understand the separation of the electron and ion perturbations. It is obvious that there will be breakdowns in this simple model; nonetheless, this simple model has proved quite serviceable for many years.

In the absence of perturbations the line shape spectrum, $\Phi(\omega)$, will be related to the sum over the dipole radiative transitions weighted by the density of atomic states ρ:

$$\Phi(\omega) \sim \sum \rho_\alpha \, d_{\alpha\beta} \cdot d_{\alpha\beta} \, \delta(\omega - \omega_{\alpha\beta}) \qquad (3)$$

where the atomic dipole matrix element between state α and β, is $d_{\alpha\beta}$, and $\omega_{\alpha\beta}$ is the transition frequency. In any interacting system, the δ-function character of the line profile will be modified by the surroundings. In a dense plasma the interaction of the plasma with the radiator will dominate. Thus, we derive, after much work, and with $<\ >$ indicating an average of the plasma,

$$\Phi(\omega) \sim \sum \rho_\alpha \int dt \, <d_{\alpha\beta}(t) \cdot d_{\alpha\beta}> e^{-i\omega t} \qquad (4)$$

Here the time development of the dipole operator is dictated by the full plasma - emitter Hamiltonian and assumptions about the initial correlations between the atom and the plasma have been made. For convenience this can be rewritten in terms of the dipole-dipole correlation function **D** as

$$\Phi(\omega) \sim \sum \rho_\alpha \int dt \, <D_{\alpha\beta}(t)> e^{-i\omega t} = \sum \rho_\alpha <D_{\alpha\beta}(\omega)> \qquad (5)$$

Introducing the concept of an ion microfield, E, which is the ion field at the emitter, and the distribution of microfields, P(E), one can derive the starting point for the discussion, i.e.,

$$\langle D(\omega)\rangle = \int dE\, P(E)\, D(\omega,E) \sim \int dE\, P(E)\, d\, [\Delta\omega - eE\cdot d - i\Gamma(\omega)]^{-1} d \qquad (6)$$

The right hand side of eq. 6 can be thought of as a convolution of a quasi-Lorentzian type profile, formed by the, frequency dependent, electron shift and width, Γ, and an ion field distribution formed by a Holtsmark-type distribution of fields. The efforts over the past three decades have been to improve / modify / extend this simple representation to overcome the various approximations inherent in the derivation.

A number of problems need to be addressed to improve the standard theory and a few will be discussed here. First, one needs to extend the treatment of the electron broadening operator into the wings, thus going beyond the impact approximation. This is called the **unified theory** and was first presented in the treatment of hydrogen lines by Smith et al. in 1969,[2] which is an evolution of formalism employed in the relaxation theories.[3] The extension to ion emitters was then achieved by Greene and Cooper.[4] Next, to incorporate the **ion dynamics** as well as the ion quasi-static broadening a novel theoretical construct was initially proposed by Frisch and Brissaud in 1971.[5] This was extended to ions over the next decade and the first ion emitter line broadening calculations were performed by Stamm et al.[6] Another outstanding problems of line broadening is the transition of the line broadening formalism from the ion quasi-static regime to the **ion impact** regime, which will occur at lower densities. The difficulty here is not to obtain the answer in the asymptotic limit, but to achieve the correct transition. This is of import since the line wings will be critically affected if the incorrect treatment of the ions is used. Research in this area is still on-going and is difficult to solve. Finally, and somewhat off the beaten track, there is the problem of the non-LTE nature of the line broadening problem. It is possible to introduce non-LTE populations into the standard formalism, but the modification of the density matrix by the plasma interactions *and* the consistent modification of the kinetics due to the line broadening are almost completely untouched problems.

The various different methods of attack on the problems of line broadening in dense plasmas can be broadly grouped into three areas. 1) The **scattering**, or collision theory, approaches[7] treat the problem as a scattering problem and determine the line *width and shift* by the thermal averaging of the detailed cross-section information. These approaches do not provide line shapes since the form of the line profile must be assumed to be a Lorentzian. 2) The **relaxation** and kinetics theories have been used since the late 1960s after the seminal work of Smith and Hooper,[8] While these approaches have been important in the elucidation of the many-body aspects of the problem recent progress has been somewhat measured. 3) The final method of attack has been the **stochastic** approach which was first presented by Frisch and Brissaud as a simple solution to the problem of introducing ion dynamical effects into the standard theory.[5] This theoretical formulations seems to have the best chance of treating the difficult problem of dense plasma line shapes.

III. THEORETICAL DEVELOPMENTS

The three areas of development for line shape theory have progressed in recent years at different paces. The most consistent has been the extensions of the standard theories made using the relaxation theoretical approach. Although the work in this area has been steady, the results have been less than exciting. The main thrusts have been to remove various of the ancillary assumptions inherent in the standard models. Thus, the inclusion of the full coulomb field, in lieu of the plasma monopole - emitter dipole interaction has been achieved.[9] In addition, the study of asymmetries due to the quadratic Stark effect, fine structure and ion quadrupole interactions have also been studied with the result that predictable, if small, modifications occur to the line shapes.[10] The work performed on the inclusion of ion dynamic corrections have also shown the trend observed in other ion dynamics formulations. That is, the line profiles at lower density are modified in the line cores by the inclusion of ion dynamics.[11] In all of these studies the relaxation and kinetics theories have, due to their simple formulation and ease of studying many-body effects, provided a fertile ground work. However, to make more dramatic advances a newer theoretical construct is needed, and that construct is the stochastic formulation which will be presented below.

Before we discuss the stochastic models a few comments should be included on recent work performed using the scattering approach to the line width and shift problem. The form of the line shape employed in the scattering calculations can be derived by taking eq. 6 for the dipole-dipole operator and using a thermally averaged cross-section to determine the width and shift function $\Gamma = <n\sigma v>$ in the line center, as

$$D_{\alpha\beta}(\omega = 0) \sim d_{\alpha\beta} [\Delta\omega - <n\sigma v>]^{-1} d_{\alpha\beta} \qquad (7)$$

The emphasis is on the calculation of the cross-sections using the best available techniques. In the past the main efforts in this area were carried on by Griem and his collaborators,[7] Wiese, Konjevic and Dimitrijevic and their collaborators using semi-classical theories,[12] and Blaha et al. using distorted wave calculations.[13] More recently the work of Seaton and his collaborators using R-matrix techniques provides this area of line width and shift calculations with a new and potentially vast, source of information.[14]

44 Spectral Line Broadening

These R-matrix calculations are being performed for a range of elements that are of importance to the study of astrophysical plasmas. The sheer number of transitions and elements alone make this a unique contribution to the field. When the level of sophistication of these calculations is taken into account the contribution is really daunting. To get an idea of the results being provided we define the width and shift function Γ as

$$\Gamma = w + id \propto Y(T) \tag{8}$$

where

$$Y(T) = \int \Omega(\varepsilon) \, e^{-(\varepsilon/T)} \, d(\varepsilon/T) \tag{9}$$

and Ω is the collision strength. In figure 1 an example of the collision strength data calculated for the 2s - 2p and 2s - 3p transitions is shown for the BeII ion. The dotted line represents the imaginary part, which contributes to the shift, while the solid line is the real part, which contributes to the width.[15] The abscissa is in units of energy normalized by Z^2.

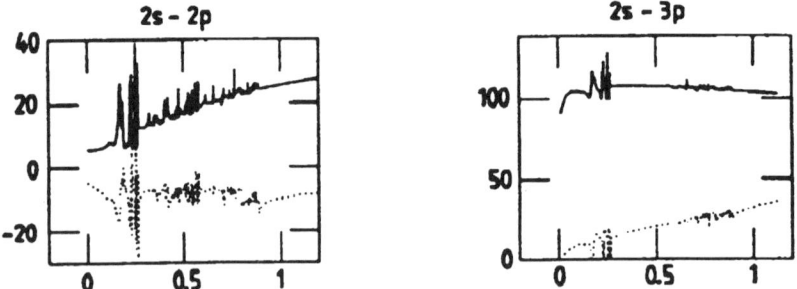

Figure 1. The results of calculations of the collision strength for two transitions in Be II. The energy units are scaled by $1/Z^2$.

Comparisons of the results of these calculation with the existing data for the widths show that there is generally good agreement. For more discussions of width data, both experimental and theoretical, see the review of line width and shift data by Wiese and Konjevic.[16] Although the calculations of Seaton are a potential major contribution to the data set for the widths and shifts of transitions, the line *shapes* of the transitions are not calculated by this techniques. Further, it is worth noting that the scattering approach as discussed here only treats the electrons and any contributions from ion-emitter interactions, which may be of importance, are neglected. In using the scattering approach to the line broadening problem the resultant shapes are constrained to be Voigt profiles, where the Lorentzian width component is provided by the calculation of the indicated in eq. 9.

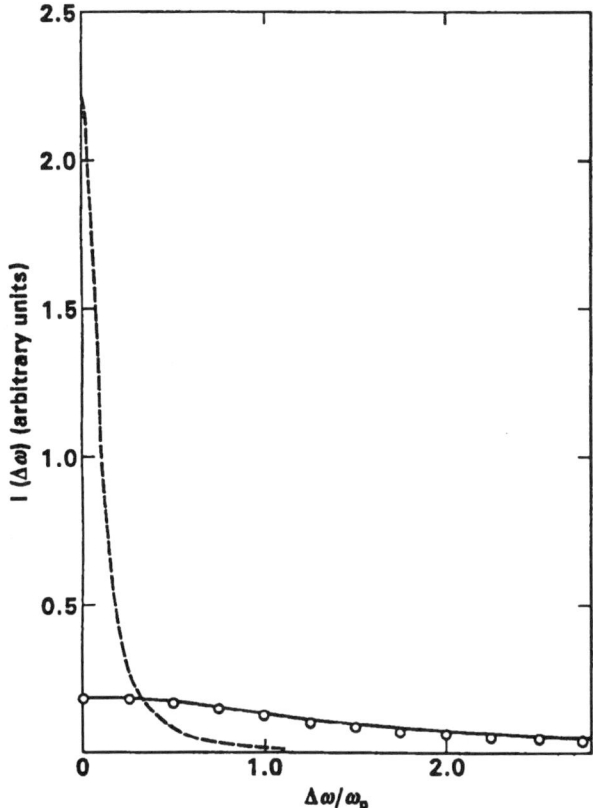

Figure 2. The Lyman α transition of hydrogenic aluminum in a plasma with a temperature of 862 eV and density of 4×10^{21} cm^{-3}. The broadening is due to protons and electrons. The solid line is the stochastic theory, the dashed line is the standard theory and the molecular dynamics simulation is indicated by the open circles.

For the generation of new theoretical work on line *shapes* the formulations using stochastic methods have produced the most interesting recent progress. The idea for this technique comes from a seminal paper by Frisch and Brissaud in 1971, which was called the model microfield method or MMM.[5] The concept is quite straight forward: since the line profile is the response of the radiator dipole to the plasma stochastic field "just" determine the effective stochastic field and the answer will fall out. This method proved very successful in answering the questions of how to include the ion dynamics and still maintain the ion quasi-static effects. Even though the choice of the functional form of the temporal evolution of the field-field correlation function was crude and the frequency dependence of the changing of the field was assumed to be proportional to the field strength, the line profiles obtained were in good agreement with existing experiments and matched simulation data.[17]

During the past few years improvements have been made in the underlying theoretical foundations. The work of Boercker at al. is a reformulation of the line shape which can assist in illustrating how the MMM can be derived in a more physical manner.[18] The inherent advantages of this work are: 1) The joint probability distribution $P(E,t|E,0)$, and thus, the field-field correlation function, $C(t)$, and the microfield distribution, $P(E)$, can all be obtained from studies in statistical mechanics. 2) The derived line profile is unified in the sense that one theory provides a line shape that is valid for ion and electrons in the core and the wings. 3) There is parallel work, which takes the same tack but uses molecular dynamics simulation to generate the field histories, that is useful for testing the analytic results.[19]

The work of Boercker derives the line shape formula as

$$\Phi(\omega) \sim \mathbf{d} \left[1 - G(\Delta\omega) \nu(\Delta\omega, E)\right]^{-1} G(\Delta\omega) \mathbf{d} \tag{10}$$

where

$$G(\Delta\omega) \sim \int dE\, P(E) \left[\Delta\omega + \nu(\Delta\omega, E) - e\mathbf{E}\cdot\mathbf{d} - i\Gamma(\omega)\right]^{-1} \tag{11}$$

Here the form of the line shape is quite simple and the $G(\Delta\omega)$ is the same line shape function as in the standard theory with the addition of the $\nu(\Delta\omega, E)$, see eq. 6. In the limit that the ν goes to zero, which implies that the field E does not change, the standard theory is recaptured. Further, the impact limit is obtained in the limit of large ν.

At the crux of the theoretical development of the stochastic theory, in the form represented by eqs 10 and 11, is the determination of the $\nu(\Delta\omega, E)$ which is the rate of change of the field at the emitter due to the plasma. The first work using the stochastic model, which was devoted to the case of neutral emitters, i.e, the MMM, guessed that the ν could be modeled as a function that was constant at a given value of the field for a time that was inversely proportional to the field. In this way the strongest fields last for the shortest times. In addition, the form for the field-field correlation function is assumed to correspond to the low density limit.[20] In the formalism of Boercker et al., which concerns the case of ionic emitters, the determination of the $\nu(\Delta\omega, E)$ is placed centrally in the development. The simplest approximation that can be *derived* is that ν is constant and related to the self-diffusion coefficient. (Note that this constant ν was modeled in the original MMM.) This provides a working model for the line shape and results are obtained for hydrogenic emitters and compared to the molecular dynamics simulations, as well as the standard theory. In general the results are impressive. In particular, the results for the Lyman α line of hydrogenic aluminum in a plasma of 862 eV and 4×10^{21} cm^{-3} is shown in figure 2. The comparison shows a large difference in the line core due to ion dynamics between the standard theory and the stochastic theory. Further, for this case the molecular dynamics simulations are in excellent agreement. Since there are no detailed line shape experiments of this type of line profile, comparison with simulation offers the best possible tests of the theory. Note that in experiments the Doppler and fine-structure, not includedn here, would be important.

The need for extensions of the standard theory also arise from another direction. That is, there are demands being placed on the calculation of line shapes relevant for all types of spectral synthesis. The need for detailed multi-frequency opacities for low to mid Z astrophysical plasmas arises from the new developments in the synthesis of opacity.[21] Further, the theory of high Z opacities has recently undergone a transformation with the work on unresolved transition arrays (UTA) and super transition arrays.[22] Although the UTA is composed of myriad line transitions the shape of the UTA has never been theoretically investigated since the number of calculations is prohibitive. Finally, demands arise from the needs of non-LTE spectral synthesis. In the non-LTE area the kinetics modeling of x-ray lasers,[23] generalized L-shell spectrum,[24] K-shell diagnostics,[25] and the newer generation of detailed radiative transfer codes treating hundreds of lines at one time,[26] all require spectral line profiles. All of these requirements have arisen since the advent of large computer and the emphasis is on speed with reasonable accuracy.

Advances in the computational aspects of the generation of line shapes come from two main sources. First is the work carried out at the Université de Provence in Marseille by Talin, Calisti, Khelfaoui and Stamm in conjunction with scientists at LLNL.[27] The main thrust of this work is to improve the speed of the standard theories, as extended to be unified, so that a line shape calculation can be performed for an arbitrarily complex emitter. Further, improvements are sought to extend the formalism so that the predictions for the line shapes will be correct for the largest range of Z, T_e and N_e possible. Second is the work carried on at the University of Florida and NIST by Hooper, Woltz and their collaborators.[28] The effort again is to calculate the line shapes for arbitrary emitters but here the theoretical development is along the lines of an extension of the relaxation theories.

The type of calculations that have been performed range from the complete line spectrum from the n = 2 to n = 3 or 4 transitions for neon-like ions,[29] to the calculation of line series, such as the n = 2 to 3, 4 and 5 in lithium-like ions.[30] In these calculations user is required to provide the atomic data, which consists of the energy levels and the dipole matrix elements of the ion to be studied. After these inputs are specified the user can select the line transitions of interest for the specific application at hand.

As an example of these calculations figures 3 and 4 shows line shapes for the helium-like argon spectrum arising from the 1s2l - 2l2l' transitions, which form satellites on the low energy side of the hydrogenic Lyman α transition.[28] The calculations are shown for a plasma with a temperature of 1000 eV and a density of 10^{23} cm^{-3}. The importance of this example is that size of the problem is large, with about 300 nlj states – requiring the diagonalization of matrices on the order of 2000. In the fig. 3 the effects of the electron broadening, the autoionization rates, radiative decay rates and Doppler effect are shown. In fig. 4 the effects of the quasi-static ions is also included to give the final line profile. The contributions of the various sources of broadening can thus be observed.

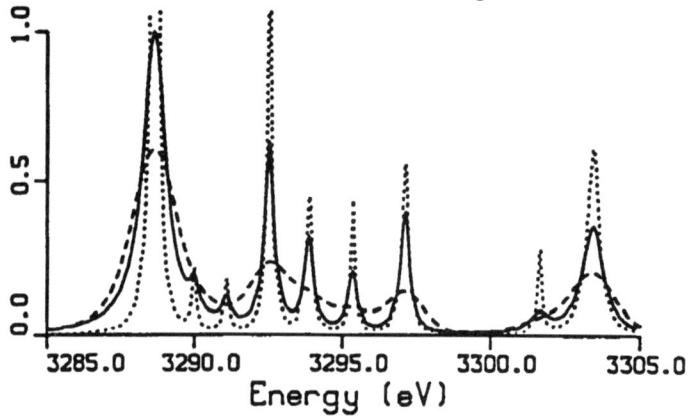

Figure 3. The He-like Ar 1s2l - 2l2l' spectrum from a plasma with T_e = 1000 eV and $N_e = 10^{21}$ cm^{-3}. The dotted line is the contributions of the electrons only, the solid line represents the additional effects of the autoionization and radiative decay rates, while the dashed line includes the Doppler convolution also. Note that for the electron only case the lowest energy feature goes off scale.

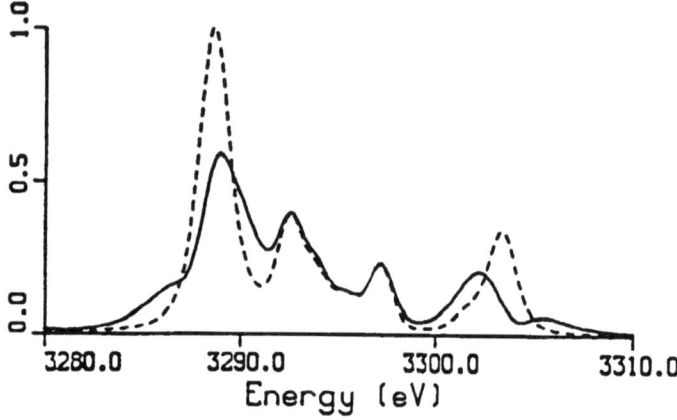

Figure 4. As fig. 3 with the solid line representing the electrons, autoionization & radiative decay rates and the Doppler effect, while the dashed line includes the further affects of the ion quasi-static fields.

In addition to the speed of the computations, work is ongoing to include a wider range of effects. There is a need to evaluate the ion dynamic corrections to the standard theory; but, the methods to perform ion dynamic calculations are, so far, prohibitively expensive. One possibility would be the use of the stochastic theory as discussed above. This option is not available yet and an alternative approach is being developed by Calisti et al.[31] This latter method of including the ion dynamic corrections is based on the idea that if the ion field fluctuates during the emission process, the emission can switch from one component to another. This may be viewed as line mixing and, in a model where the mixing is performed according to a stationary Markovian process,[32] mixing by line pairs is calculated. The critical ingredient in the model is the fluctuation parameter ν, which is determined by the ion microfield fluctuations. Note that this ν and the one needed for the stochastic theories as outlined above are essentially the same. In practice molecular dynamics calculations of the field-field correlation function have been performed and the inverse of the correlation time τ is identified as ν.

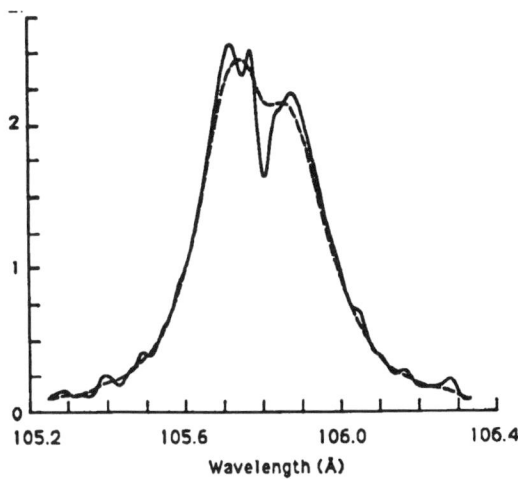

Figure 5. The lithium-like aluminum 5f-3d transition in a plasma with temperature of 200 eV and a density of 10^{20} cm^{-3}. The solid line represents the results of the standard theory while the dashed line is the ion dynamics approach using the line mixing approach.

In figure 5 the results of a calculation of the lithium-like aluminum 5f - 3d transition at a density of 10^{20} cm^{-3} and a temperature of 200 eV is shown. Figure 5 shows a comparison of the effects for the standard theory and one with the dynamic ions included. The effects of the line mixing - and ion dynamics in general - are two fold. First, the line mixing reduces the lifetimes of the homogeneous components, resulting in an increased width of order v. Second, line mixing tends to move those close components together, which, depending on the placement of the components, may cause a narrowing or broadening of the complete profile. In fig. 5 the the dip seen in the standard theory is smeared out but the half width is left unchanged. Finally, note that the increase in the computational time for the ion dynamic calculation is only a factor of two greater than for the standard theory. This is a very promising approach, if it turns out that these ion dynamic predictions are accurate over a wide range of conditions.

IV. EXPERIMENTS

The number of experiments that are devoted to line shape measurements, as opposed to those that use line shapes as a diagnostic, is small. The main proponents of line shape work in the high density plasma research are Hooper in consort with experimenters at the National User Laser Facility (NULF). The NULF experiments are devoted mainly to attempting to access regions of extremely high density, in excess of 10^{23} cm^{-3}, which usually yields interesting data but with insufficient accuracy to critically affect the theoretical development.[33]

For lower density plasma experiments the work at the University of Maryland lead by Griem and at Ruhr Universitat lead by Kunze is directly relevant. These efforts on plasmas of interest to x-ray laser production and general studies of laser produced plasmas have led to a number of contributions on plasma line shapes in recent years.[34] However, the difficulty in producing a well characterized plasma has kept these efforts modest. On a slightly different tack there are more examples of the studies of line widths. These are have been carried out by NIST and Institute of Physics in Yugoslavia, in addition to the above institutions.[35]

V. FUTURE WORK

The future is difficult to predict but it seems that there are two clear trends which can safely be commented upon. First, there is the outstanding problem of the far line wings. This problem has been ignored in the previous discussion because very little has been done about it. Second, there is the clear road toward a practical ion dynamics theory.

The problem with line wings, and here the far line wings are of interest, is important, for example, to the study of opacity. The importance arises from the fact that most energy transport occurs in the low opacity regions, i.e., the "windows" of the opacity. In astrophysical plasmas, where there are relatively small contributions from high Z ions, the spectral character is not dominated by UTAs and thus individual transitions, albeit myriad, contribute a large part of the opacity. Since the windows are dominated by the wings of line transitions, and continua, the detailed line shapes in the far line wings become important.

The problem with calculating the line shape of the far wings arises from the fact that strong close collisions, which will dominate, have numerous competing channels – which any good scattering calculation includes. On the one hand, for electrons interacting with ionic emitters the processes of recombination, electron capture and ionization become important as do the other inelastic processes. For the ions, on the other hand, there are the possibilities of quasi-molecular satellites which would demand detailed calculations.

The line shape formalisms used here assume that the emitter system is initially independent of the plasma. In effect this excludes a simple rigorous method to obtain a solution to the problem. However, a scattering formalism does not have a method to allow the appropriate plasma effects to be included. Both are needed to get the correct line wing contributions. This problem has been alluded to and commented on before;[36] but, the solution for the far wings has never been actively sought.

The problem of far line wings is related to the arbitrary, but serviceable, separation that is made in performing kinetics calculations. A simple example arises in the study of gain in possible x-ray laser systems. In the kinetics equations the rates between levels provide the mechanism for populating the states, the gain however requires both the populations and the line shape. The line broadening processes relevant to the transitions can substantially alter the lifetimes of the states of interest and change the gain predictions dramatically. It is not difficult to find a case where the separation of these two closely related phenomena, kinetics and line broadening, creates inconsistencies. This is a problem for the future that currently has no clear solution.

The second prediction for the future is on somewhat firmer ground. The work to be done on the formulation of the stochastic theories of line broadening requires information on the conditional probability function, $P(E,t|E,0)$, for the field at the emitter. The $v(\Delta\omega,E)$ is essentially the rate of change of the $P(E,t|E,0)$ and the field-field correlation function, $<E(t)\cdot E(0)>$, is the second moment of this distribution. Thus, if one could generate the $P(E,t|E,0)$ for a charge point there would be a good chance to determine the parameters needed by the stochastic theories.

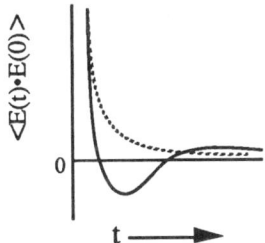

Figure 6. A schematic of the temporal behavior of the field-field correlation function. The solid line represents the correct behavior and the dashed line represents the approximate behavior assumed in the MMM formulations.

The original guess for the behavior of the $<E(t) \cdot E(0)>$ made in the MMM formulation for the neutral case was that correlation function should fit the long and short time limits. An extension of the MMM to the case of charged emitters would lead to a function that behaves like the dashed line in figure 6. In fig. 6 the temporal dependence of the correct $<E(t) \cdot E(0)>$ is shown as the solid line. Noting that it can be rigorously proved that the integral over time of the correlation must be zero for a charged point, it is seen that the MMM will not match the real functional dependence. This should be coupled to the fact that Boercker has shown that when $<E(t) \cdot E(0)>$ is modeled poorly, the line shapes predicted by the stochastic methods are in poor agreement with simulations.[37] Further, the disagreement becomes worse as the strong coupling parameter becomes larger. Thus, for a complete and accurate theory, valid over the full range of temperature, density and Z one would need to obtain better predictions of the $<E(t) \cdot E(0)>$ and the $v(\Delta\omega, E)$. These, in turn, require the $P(E, t | E, 0)$ to be obtained for a fuller range of plasma conditions.

There is a substantial effort in studies of the statistical mechanics of plasmas to determine the $P(E, t | E, 0)$ and recent work by Alastuey et al.,[38] on the electric field dynamics at a neutral point provides line shape theorists with hope that these methods can be extended to charged points. The results of a study of the type Alastuey et al. carried out would be that the stochastic methods for line broadening would have field histories from an analytic theory. One could then do molecular dynamics simulations of the line broadening with these field histories, while the results could then be compared to an analytic stochastic method, e.g., Boercker et al., where the necessary properties $<E(t) \cdot E(0)>$ and the $v(\Delta\omega, E)$ are derived from the electric field dynamics at a charged point. This would provide a major step forward in the quest for a completely unified line shape theory.

VI. SOURCES FOR LINE SHAPES

In this section a brief accounting will be given of those resources available for line shape information. Although this list is not exhaustive, and in fact is quite limited, these form a reasonable starting point for any researcher.

For general information on all aspects of line broadening both theoretical and experimental the most obvious contact is Professor H. Griem of the Laboratory for Plasma Research, University of Maryland in College Park. His group over the years has been involved in every aspect of line broadening and remains a central source of information. In the same vein the Data Center on Atomic Line Shifts and Widths at the National Institute of Science and Technology in Gaithersburg, Maryland headed by Dr. W. Wiese collects and evaluates all the width and shift information possible. In addition, Dr. Wiese and his colleagues are involved in determining scaling formulae for widths and shifts which can greatly extend the amount of information available.

For the computation of line shapes for arbitrarily complex structures, essentially within the constraints of the standard theory, one has two sources. One is the group working at the University of Florida at Gainesville under the supervision of Professor C. Hooper, and the other at the Université de Provence in Marseille - St. Jerome Center under the leadership of Drs B. Talin and R. Stamm. Both of these groups is interested in new and challenging problems in the area of charged emitter line broadening.

Finally, there are codes for the simplest cases of hydrogenic Lyman and Balmer series line shapes, helium-like $1s\ ^1S - 1snp\ ^1P$ line shapes and the lithium-like series $1s^22l - 1s^2nl'$ where l is 0 (s) or 1 (p) and l' runs the full range for principal quantum number n, i.e., $0 \le l \le n-1$. These codes are available from R. Lee upon request. They run on Macintosh computers and other 32 bit desk top computers in a reasonably short time. In addition, for hydrogenic ions there has been much work. Some recent examples can be found in the tables for the Balmer, Paschen and Pickering series of HeII calculated in the unified theory for astrophysical condtions by T. Schoning and K. Butler,[39] and also the work by Stehlé employing an extension of the MMM approach to ion dynamics.[40]

VII. BIBLIOGRAPHY

A few papers are listed below that may be of particular interest, which have been used in composing this review. Some, but not all, are reproduced in the references. Many of these contributions have been provided by individuals kind enough to send papers, preprints and drafts of papers on topics related to the study of spectral line broadening of ion lines.

Radiative Transfer Effects:

- "Spectral line broadening from induced radiative processes in soft x-ray laser amplifiers", – A study of the effects ASE on line shapes – H. Griem and J. Moreno preprint UMLPR 91-039 submitted to Phys. Rev A
- "Nonlinear Interference Effects in Partial Redistribution of Radiation in Dense Plasmas", A. Demura et al. an Invited paper presented to the 10th International Spectral Line Shape Conference page 227 in "Spectral Line Shapes" 6, edited by L. Frommhold and J. Keto (American Institute of Physics, New York, 1990)
- "Radiative transfer and transverse inhomogeneity effects in spectral lines emitted from laser plasmas", C. Bousquet, J. Grumberg, E. Leboucher-Dalimier, H. Nguyen and A. Poquerusse, J. Phys. B **23**, 1783 (1990)
- "Asymmetric repumping of the Lyman α components of hydrogen-like ions in a dense expanding plasma", F. Rosmej, A. Schulz, K. Koshelev and H.-J. Kunze, J.Q.R.S.T. **44**, 559 (1990)

Experimental Results:

- "Measurements of line broadening of BV H_α and L_δ in a laser-produced plasma" J. Wang, H. Griem, Y. Huang and F. Bottcher, preprint
- " Regularities in Experimental Stark Shifts", W. Wiese and N. Konjevic, submitted J.Q.R.S.T. 1991
- "Are Einstein's Transition Probabilities for Spontaneous Emission Atomic Constants?" H. Griem, Y. Huang, J. Wang and J. Moreno, Phys. Fluids B **3**, 2430 (1990)
- "Experimental Stark Widths and Shifts for Spectral Lines of Neutral and Ionized Atoms (A Critical Review of Selected Data for the Period 1983 through 1988)" N. Konjevic and W. Wiese, Reprint #402 J. of Phys. and Chem. Ref. Data **19**, 1307 (1990)
- "Measurements of Stark broadening of BV and NV n = 6 to 7 and 5 to 6 transitions in a laser produced plasma", Y. Huang, F. Bottcher, J. Wang and H. Griem, Phys. Rev. **A42**, 2322 (1990)
- "Measurement and Interpretation of Spontaneous Line Emission from a Common Upper Level in Laser-Produced Carbon Plasmas" – broadened line profiles may be the solution of a puzzling result – Y. Huang, J. Wang, J. Moreno and H. Griem, Phys. Rev. Lett. **A65**, 1757, (1990)
- "Stark shift of the HeII P_α line in a dense plasma", A. Gawron, J. Hey, X. Xu, and H.-J. Kunze, Phys. Rev. **A40**, 7150 (1989)
- "Production of hot near-solid-density plasma by electron energy transport in a laser-produced plasma" – line shapes are used as a diagnostic – G. Tallents, M. Key, P. Norreys, D. Brown, J. Dunn, and H. Baldis, Phys. Rev. **A40**, 2857 (1989)
- "Stark broadening of Ar IV lines in a dense plasma" J. Hey, A. Gawron, X. Xu, P. Breger and H.-J. Kunze, J. Phys. B **22**, 241 (1989)
- "Dependence of the Stark Broadening on the emitter charge for the 3s-3p transitions of Li-Like ions", F. Bottcher, P. Breger, J. Hey and H.-J. Kunze, Phys. Rev. **A38**, 2690 (1988)

Theoretical Results:

- "Time-dependent Statistical Properties of the Electric Microfield Seen By a Neutral Radiator" – This is a calculation of the joint probability function for the field at a neutral point, which if it could be performed for a charged point would greatly assist the stochastic theoretical development. – A. Alastuey, J. Lebowitz and D. Levesque, Phys. Rev **A43**, 2670 (1991)
- "Dynamics of Electric Field in Strongly Coupled Plasmas", – Limiting cases for the field-field correlation function which is essential to the stochastic theories.– J. Dufty and L. Zogaib, page 535 in "Strongly Coupled Plasma Physics" edited by S. Ichimaru (Elsevier Science Publishers, 1990)
- "Atomic data for opacity calculation: XIII. Line profiles for transitions in hydrogenic ions" – A numerical method for getting rapid hydrogenic ion

calculations within the context of the standard theory. There is an excellent discussion in the appendices of the numerical aspects of the formulation. – M. Seaton, J. Phys. B **23**, 3255 (1990).

• "Atomic data for opacity calculation: I through XII" – A series of paper which discuss calculation using R-Matrix formulation for the cross-sections. The use of the best available method and the sheer volume of the calculational scope in ion stage and elements will substantial increase the eventual shift and width data base – M. Seaton, J. Phys. B **20** to **22**.

ACKNOWLEDGEMENTS

I would like to take this opportunity to thank all the people who have sent me copies of their work and provided comments. I would also like to thank all the individuals at LLNL and the Université de Provence for helping me understand, as much as possible, some of the subtleties in the field of line broadening.

VIII. REFERENCES

1. M. Baranger, Phys Rev **111**, 481 & 494 (1958) and **112**, 855 (1958); A. Kolb and H. Griem, Phys. Rev. **111**, 514 (1958)
2. E. Smith, J. Cooper and C. Vidal, Phys. Rev. **185**, 140 (1969); C. Vidal, J, Cooper and E. Smith, J.Q.S.R.T. **10**, 1011 (1970) & **11**, 263 (1971) and for tabulations of some neutral hydrogen Lyman transitions "Unified Theory Calculations of Stark Broadened Hydrogen Lines Including Lower State Interactions" National Bureau of Standards Monograph #120 (Washington D.C., U.S. Dept of Commerce, 1971)
3. See E. Smith and C. Hooper, Phys. Rev **157**, 126 (1967) and E. Smith, Phys. Rev **166**, 102 (1968)
4. R. Greene and J. Cooper, J.Q.R.S.T. **15**, 1025, 1037, & 1045 (1975)
5. U. Frisch and A. Brissaud, J.Q.S.R.T. **11**, 1753 & 1767 (1971)
6. R. Stamm, Y. Botzanowski, V. Kaftandjian, B. Talin and E. Smith, Phys. Rev Lett. **52**, 2217 (1984) & **54**, 2170 (1985); and culminating in the thorough calculations of R. Stamm, B. Talin, E. Pollock and C. Iglesias, Phys. Rev. **A34**, 4144 (1986)
7. H. Griem "Spectral Line Broadening by Plasmas", chapter II and appendix V (Academic Press, New York, 1974)
8. E. W. Smith and C. F. Hooper, Phys. Rev. **157**, 126 (1967) and **166**, 102 (1968) are the first of a series of papers using the relaxation theory to investigate the statistical mechanical problems of line broadened.

9. J. O'Brien and C. Hooper, J.Q.S.R.T. **14**, 479 (1974); R. Tighe and C. Hooper Phys. Rev **A14**, 1514 (1976); and culminating in L. Woltz and C. Hooper, Phys. Rev. **A38**, 4766 (1988)
10. R. Joyce, L. Woltz and C. Hooper, Phys. Rev. **A35**, 2228 (1987) – ion quadrupole and asymmetries; L. Woltz and C. Hooper in "Proceedings of the 2nd International Conference on Radiative Properties of Hot, Dense Matter" 476, (World Scientific, Singapore, 1987) – quadratic Stark effect and fine structure
11. R. Greene, J. Phys. B **15**, 1831 (1982); D. Oza, R. Greene and D. Kelleher, Phys. Rev. **A34**, 4519 (1986) – Note that the theoretical work described here is a hybrid, since it requires simulation results to determine the fluctuations in the ion field.
12. See, for example, N. Konjevic and W. Wiese, J. Phys. and Chem. Ref. Data **19**, 1307 (1990); M. Dimitrijevic and N. Konjevic J.Q.S.R.T. **24**,427 (1980) and the work on experimental scaling of J. Puric et al., Phys Rev **A35**, 2111 (1987), **A36**, 3957 (1987) & **A37**, 498 (1988). For recent calculations on He-like argon, including shifts, see H. Griem, M. Blaha and P. Kepple, Phys. Rev. **A41**, 5600 (1990). In addition, the experimental scaling work of J. Hey and P. Breger should be noted, J.Q.S.R.T. **23**, 311 (1980), **24**, 349 & 427 (1980) and S. Afr. J. Phys. **5**, 111 (1982)
13. See the distorted wave calculations as applied in H. Griem, M. Blaha and P. Kepple, Phys. Rev. **A19**, 2421 (1979)
14. See the series of papers on atomic data calculations, for example, M. Seaton, J. Phys. B **20**, 6431 (1987)
15. M. Seaton, J. Phys. B **21**, 3033 (1988)
16. N. Konjevic and W. Wiese, J. Phys. and Chem. Ref. Data **19**, 1307 (1990)
17. A. Brissaud, C. Goldbach, J. Léorat, A. Mazure, and G. Nollez, J. Phys. B **9**, 1129 (1976) ;
18. D. Boercker, C. Iglesias and J. Dufty, Phys. Rev. **A36**, 2254 (1987)
19. R. Stamm, B. Talin, E. Pollock and C. Iglesias, Phys. Rev. **A34**, 4144 (1986)
20. A. Brissaud, C. Goldbach, J. Léorat, A. Mazure and G Nollez, J. Phys. B **17**,1477 (1984); U. Frisch and A. Brissaud, J.Q.S.R.T. **11**, 1767 (1971)
21. See for example, C. Iglesias, F. Rogers and B. Wilson, Ap. J. **360**, 221 (1990)
22. For a review of unresolved transition arrays see C. Bauche-Arnoult, J. Bauche and M. Klapisch, Adv. At. Mol. Phys. **23**, 131 (1988); for the super transition arrays see A. Bar-Shalom, J. Oreg, W. Goldstein, D. Shvarts and A. Zigler, Phys. Rev **A40**, 3183 (1989)
23. See the reviews in the book by R. Elton "X-ray Lasers" (Academic Press, Boston, 1990)
24. For example, W. Goldstein, R. Walling, J. Bailey, M. Chen, R. Fortner, M. Klapisch, T. Phillips and R. Stewart, Phys. Rev. Lett. **58**, 2300 (1987)

25. See for example the examples in R. W. Lee, B. L. Whitten, and R. E. Strout, J. Q.S.R.T. **32**, 91 (1984) and B. d'Etat, J. Grumberg, E. Leboucher, H. Nguyen and A. Poquerusse, J. Phys. B **20**, 1733 (1987)
26. R. Klein, J. Castor, A. Greenbaum, D. Taylor and P. Dykema, J.Q.S.R.T. **41**, 199 (1989); and D. Band, R. Klein, J. Castor and J. Nash, Ap. J. **362**, 90 (1990)
27. A. Calisti, R. Stamm, B. Talin, and R. Lee, Phys. Rev. **A42**, 5433 (1990)
28. L. Woltz and C. F. Hooper, Phys. Rev. **A38**, 4766 (1988); see also Phys. Rev. **A44**, 1281(1991)
29. C. Keane, B. Hammel, A. Osterheld, L. Suter and R. Lee, Rev. Sci. Instrum. **61**, 1 (1990)
30. Private communication from the author to himself
31. A. Calisti, F. Khelfaoui, R. Stamm and B. Talin in "The Proceedings of the Conference on Radiative Properties of Hot Dense Matter", edited by W. Goldstein et al. (World Scientific, Hong Kong 1991)
32. B. Talin, Y. Botzanowsky, C. Calmes and L. Klein, J. Phys. B **16**, 2313 (1982)
33. N. Delamater, C. Hooper, R. Joyce, L. Woltz, N. Ceglio, R. Kauffman, R. Lee and M. Richardson, Phys. Rev. **A31**, 2460 (1985)
34. Experiments on line shapes: Y. Huang, F. Bottcher, J. Wang and H. Griem, Phys. Rev **A42**, 2322 (1990); A. Gawron, S. Maurmann, F. Bottcher, A. Meckler and H-J Kunze, Phys. Rev **A38**, 4737 (1988)
35. F. Bottcher, P. Breger, J. Hey, and H.-J. Kunze, Phys. Rev. **A38**, 2690 (1988) and J. Hey A. Gawron, X. Xu, P. Breger and H.-J. Kunze, J. Phys. B **22**, 241 (1989); See also the compendium published by the Data Center on Atomic Line Shapes and Shifts at NIST, e.g., N. Konjevic and W. Wiese, J. Phys Chem. Ref. Data **19**, 1307 (1990)
36. C. Stehlé, J. Q. S. R. T. **44**, 135 (1990); A. Demura, A. Anufrienko, A. Godunov, Y. Zemtsov, V. Lisitsa, A. Starostin, M. Taran and V. Schipakov, page 227 in "Spectral Line Shapes" **6**, edited by L. Frommhold and J. Keto (American Institute of Physics, New York, 1990); A. Pradhan, D. Norcross and D. Hummer, Ap. J. **246**, 1031 (1981)
37. D. Boercker in "Spectral Line Shapes" **5**, 73, edited by J. Szudy (Ossolineum, Warszawa, 1989)
38. A. Alastuey, J. Lebowitz and D. Levesque, Phys. Rev **A43**, 2670 (1991)
39. T. Schoning and K. Butler, Astron. Astrophys. Suppl. Ser. **78**, 51 (1989)
40. C. Stehlé in "The Proceedings of the Conference on Radiative Properties of Hot Dense Matter", edited by W. Goldstein et al. (World Scientific, Hong Kong 1991)

X-RAY AND OPTICAL DIAGNOSTICS OF FEMTOSECOND LASER-PRODUCED PLASMAS

P. Audebert, J.P. Geindre, J.C. Gauthier
Laboratoire de Physique des Milieux Ionisés
Ecole Polytechnique, 91128 Palaiseau, France

R. Benattar
Laboratoire d'Utilisation des Lasers Intenses
Ecole Polytechnique, 91128 Palaiseau, France

J.P. Chambaret, A. Mysyrowicz, A. Antonetti
Laboratoire d'Optique Appliquée, ENSTA-X,
Batterie de l'Yvette, 91120 Palaiseau, France

ABSTRACT

Layered targets of aluminum and fused silica have been irradiated by 80 fs prepulse free laser pulses at 10^{17} W/cm^2 laser irradiance. Time-integrated x-ray spectra obtained in the 0.7-0.9 nm wavelength range show both resonance lines from Al^{11+} ions and Kα lines from Al^{0+}-Al^{5+} ions. Conversion efficiency measurements give x-ray intensities of a few TW/cm^2. A detailed analysis of Kα emission points out the role of energetic electrons in the interaction physics. Based on a detailed comparison of experiments with hydrodynamic simulations, it is concluded that Kα emission is not longer than the laser pulse duration.

INTRODUCTION

The development of lasers delivering ultra-short pulses at the multigigawatt power level opens new perspectives for the study of laser-matter interaction at very high intensities. When an intense ultra-short pulse laser is focused onto the surface of a solid, a high temperature, x-ray emitting plasma is produced. Laser energy heats up the solid target material over a length comparable to the skin depth. Due to the short laser pulse duration, this leads to rapid ionization before any significant hydrodynamic motion of the target occurs. These plasmas look very promising for the generation of sub-picosecond x-rays in the keV range. Several groups[1-6] have recently begun to explore the physics of laser-heating solid density targets because the availability of such x-ray pulses will have significant impact in many areas requiring time-resolved measurements in plasma and solid state physics.

We report here the first experiments using K shell emission to investigate the laser energy deposition and transport mechanisms of femtosecond laser-produced plasmas. High resolution x-ray spectra of Kα radiation from low-charge aluminum ions points to the fact that energetic electrons play a significant role in the laser energy transport. Measurements of Al^{11+} and Si^{12+} line intensities as a function of aluminum thickness in aluminum and fused silica layered targets give a penetration depth in the range 100-150 nm for a laser intensity of 10^{17} W/cm^2.

Using hydrodynamic simulations where the heat transport is calculated by non-local heat flux[7] methods and the hot electron energy deposition by a flux-limited diffusion technique[8], we infer target heating to distances about 5 times larger than the non-linear heat wave penetration depth, in agreement with the layered targets data. The code preheat temperature is consistent with the value deduced from the experimentally determined average charge state of the preheat plasma through the high-temperature Fermi model of an average ion in a cell[9]. In the simulations, the laser energy deposition is evaluated by solving the Fresnel equations[10] with the refractive index given by a solid state model at each time step. Our work suggest an original method to produce subpicosecond keV x-ray pulses from laser-produced plasmas using K-shell emission from nonequilibrium electrons.

EXPERIMENTAL SYSTEM AND OPTICAL DIAGNOSTICS

We use a CPM dye laser that delivers an optical pulse of 80 fs duration at 620 nm wavelength with a maximum energy per pulse of 1.5 mJ. The repetition rate of the laser is 20 Hz. The laser beam is focused at normal incidence with a f/3.5 lens onto a solid target so that 80% of the laser energy is located in a 10 μm^2 focal spot. The focal spot size is measured by an equivalent image-plane technique. These experimental conditions produce a typical laser intensity of 10^{17} W/cm^2 on target. The laser pulse temporal shape is analyzed by a third-order induced-grating autocorrelator[11] with a dynamic range of about 10^5. Results give a pulse duration of 80 fs (sech^2) at full laser intensity. Energy measurements with calibrated diodes show that the femtosecond pulse is superimposed on a 3 ns FWHM pedestal of amplified spontaneous emission (ASE) which contains 4% of the total energy. The femtosecond pulse starts 1.5 ns before the peak of the ASE under normal conditions. The irradiance in the ASE pulse being of the order of a few 10^{10} W/cm^2, it produces a plasma with which the high intensity pulse interacts. Schlieren and shadowgraphy measurements of the plasma created by the ASE prepulse show that the critical density of this plasma is located at less than 1 μm from the solid target surface[12].

Measurements of the energy reflected by the target indicate that 20 to 30% of the laser energy is absorbed. The targets used in these experiments are thin layers of aluminum of different thicknesses deposited on a fused silica substrate. The target is mounted on a X-Y motorized translation unit controlled by a micro-computer in order to expose a fresh surface to the laser at each shot.

The x-ray emission of the plasma is analyzed by a Von Hamos spectrograph equipped with a pentaerythritol (PET) crystal having a 10 cm curvature radius. We use this geometry for its high luminosity: x-rays falling on the 5 cm width of the crystal are focused to 300 µm on the film. The data are recorded on SB392 film filtered by a 20 µm thick Be foil in order to eliminate visible light exposure. The spectrum is then digitized and processed on a micro-computer. Film density is converted to emissivity (keV/keV/steradian) by using the known crystal reflectivity and the calculated transmission of the Be filter. We present in Fig.1 a typical spectrum obtained with a target consisting of a 300 nm aluminum layer deposited on a fused silica substrate after 15000 shots.

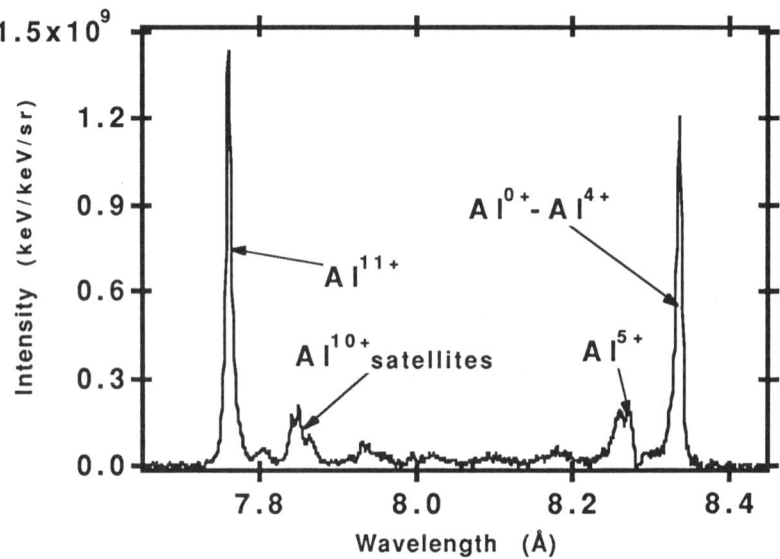

Fig 1. *X-ray spectrum of aluminum obtained with the Von Hamos spectrograph (15000 shots).*

Kα lines of aluminum from Al^{0+} to Al^{5+} are obviously prominent in the spectrum. Emissions from Al^{6+} to Al^{9+} are barely visible. The emission of the 1s^2-1s2p line of the helium-like ion (Al^{11+}) at 7.75 Å and of the dielectronic satellites originating from the lithium-like ion indicate that the plasma electron temperature is a few hundred eV.

This spectrum is characteristic of the emission of a two-component plasma. The first, a high temperature plasma which emits the helium-like resonance lines and its dielectronic satellites and the second, a weakly ionized plasma heated and excited by fast electrons which produce Kα emission from Al^{0+}-Al^{4+} at 8.34 Å (the Al^{4+} line is blended by Al^{0+}-Al^{3+} lines) and from Al^{5+} line at 8.25 Å.

ENERGY PENETRATION DEPTH

To study the energy penetration depth by thermal transport through the solid we have performed a series of experiment with multilayered targets and we have measured the x-ray emission of the inner layer (fused silica) material as a function of the outer layer (aluminum) thickness.

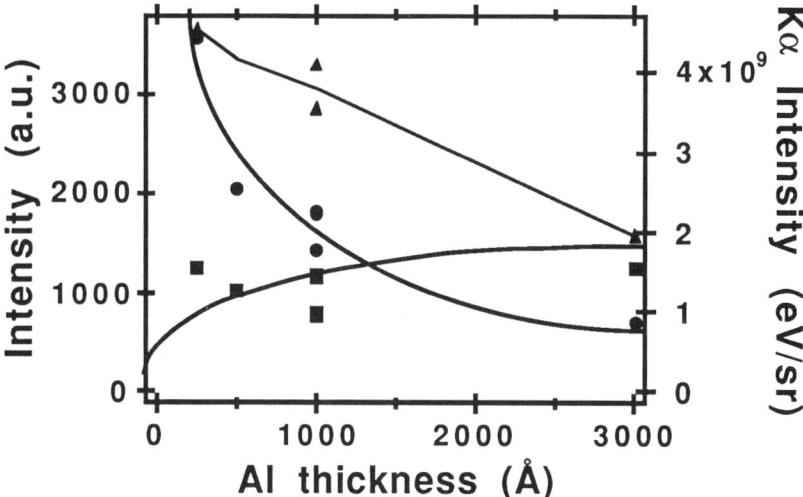

Fig. 2. *Aluminum He-like emission (squares), silicon He-like emission (circles) and silicon Kα line emission (triangles) as a function of aluminum thickness. The lines are drawn as an aid to the eye.*

We measured simultaneously the intensity of the Si^{12+} 1s2p ^1P - $1s^2$ $^1S_0$ line and of the Kα line of silicon and the intensity of the Al^{11+} 1s3p ^1P - $1s^2$ $^1S_0$ line of aluminum. Results are presented in Fig.2. The silicon emission decreases over ≈100 nm when the aluminum thickness increases. Conversely, the emission of aluminum increases. This shows that the extension of the conduction zone under our experimental conditions is of the order of 100 nm. A larger energy penetration depth of 250 nm has been obtained with a KrF laser by Zigler et.al.[13] but with a longer (600 fs) laser pulse duration.

The emission of the Kα line of silicon (triangles in Fig.2) shows a decrease of a factor of two when the aluminum thickness increases from 25 to 300 nm. For a 300 nm aluminum target (as in Fig.1) and assuming that the hot plasma thickness is only 100 nm, this implies that the hot electron energy is rather low, i.e. below 20 keV, in order to present such a high attenuation over the remaining 200 nm of the cold target[14].

ANALYSIS OF THE HOT PLASMA

We have already pointed out that the emission of the helium-like resonance lines of aluminum in the spectrum of Fig.1 is characteristic of a hot plasma with a temperature of a few hundred eV. A very significant feature of this spectrum is that there is no intercombination line ($1s2p\ ^3P - 1s^2\ ^1S_0$) at 7.807 Å. We also notice that the line profile of the $1s2p\ ^1P - 1s^2\ ^1S_0$ transition is broadened in the wings. Both features are characteristic of the emission by a high density plasma[15].

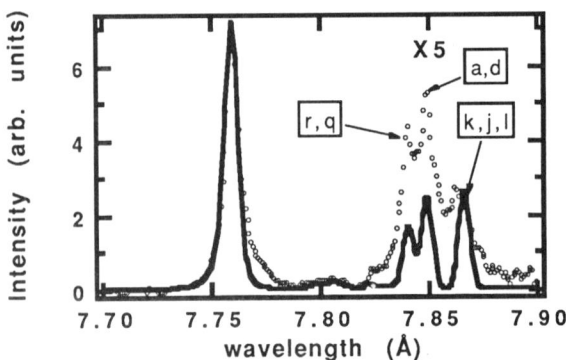

Fig. 3. *The $1s2p\ ^1P_1-1s^2\ ^1S_0$ line of the helium-like ion and the dielectronic satellites of the lithium-like ion of aluminum. Points: experiment, solid line: synthetic spectra calculated with an electronic temperature of 500 eV and an electronic density of 10^{23} cm^{-3}.*

To analyze the spectrum, we have used the code RATION[16], a computer model which produces synthetic K-shell spectra. The populations are calculated by a steady state kinetic model that includes all relevant collisional and radiative processes among a very detailed set of excited states of the lithium-, helium- and hydrogen-like ions. The microfield distribution and the shape of the resonance lines are determined using detailed calculations. Dielectronic satellite line broadening is not included in the model: only the instrumental line profile is taken into account. From this code we fit the measured resonance line profile by adjusting the temperature and the density, taking into account the instrumental spectral resolution and the plasma thickness (optical depth) that we determine from the multilayer experiment. The best fit is obtained for an electronic temperature of T_e=500 eV and an electron density of N_e=10^{23} cm^{-3}. It should be emphasized that the experimental results are space- and time-integrated so that these values should be taken as a measurement of the average plasma conditions prevailing during the laser pulse.

Comparison between the experimental and theoretical profiles is shown in Fig.3. The calculated spectrum correctly fits the center of the resonance line and the short wavelength wing. The long wavelength wing does not match because of the presence of the 2l3l' and 3l3l' satellites of the lithium-like ion which are not included in the model. Globally, the intensity of the lithium-like satellites are in good agreement with the experimental data. However the observed emissions of the satellites (a,d) and (r,q) are noticeably higher than predicted[17]. These satellites are mostly populated by inner-shell electron collisional excitation from the ground state lithium-like ion. Taking this effect into account, the discrepancy can be explained by non-stationnary effects not included in the synthetic spectra calculations which underestimate the lithium-like population.

ANALYSIS OF THE HOT ELECTRON PREHEATED PLASMA

On the longer wavelength side of the spectrum in Fig.1, we can see the 1s-2p Kα lines of Al $^{0-4+}$ ions and the 1s-2p Kα line of the Al $^{5+}$ (oxygen-like) ion. These lines are produced by the interaction of fast electrons with the solid. The total photon yield of the Kα lines found in our experiment is $Y_k = 7\pm2\ 10^9$ eV/sr, taking into account the measured attenuation of the Kα signal shown in Fig.2.

From the measured ratio of the different Kα lines, taking into account the branching ratio of the fluorescence and Auger decays[19] for all the upper levels which play a role in the Kα line formation, we have obtained the relative population of the different aluminum ions from which we have calculated the average charge Z_{eff} of the plasma. The average Z_{eff} measured by this technique is rather uncertain due to the fact that the wavelengths of the Kα lines of the aluminum ions from neutral to neon-like overlap at a wavelength of 8.34 Å and that the fluor-like line is partly blended at 8.327 Å. The Thomas-Fermi theory[9] gives an average Z_{eff} of 2.5 for neutral aluminum. We have tested the two extreme cases where the line at 8.34 Å represents the Kα emission of i) purely neutral aluminium or ii) purely neon-like aluminum: the upper and lower limit of the average Z_{eff} deduced from the experimental line ratio of the Al $^{0-4+}$, Al $^{5+}$ and Al $^{6+}$ lines is 2.99 < Z_{eff} < 3.35.

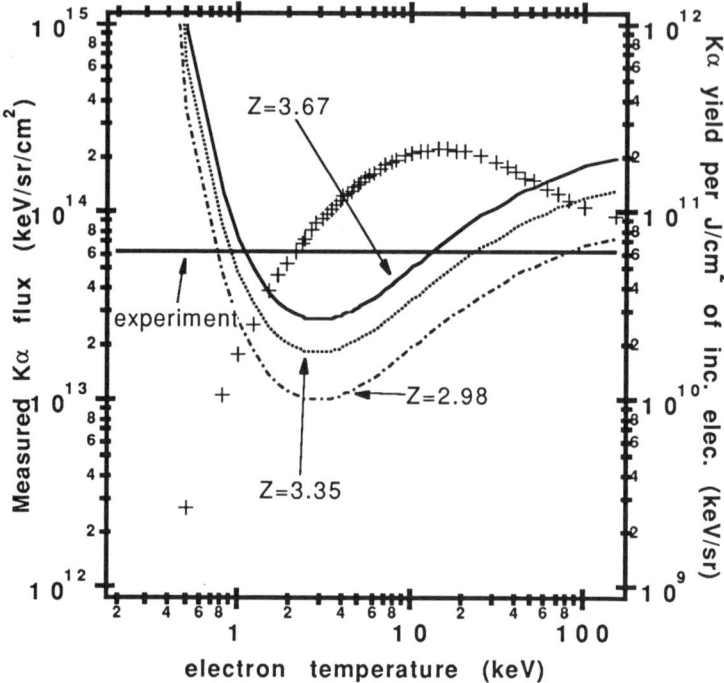

Fig. 4. *Measured Kα flux as a function of the hot electron temperature for different values of the average Z_{eff} of the target. The right scale, associated with the crosses, gives the Kα yield as a function of the hot electron temperature for an incident electron flux of 1 J/cm².*

Assuming that the fast electron preheated plasma is close to solid density, we have related the average Z_{eff} to the internal energy by using the QEOS equation of state[20]. The internal energy E_{int} of the fast electron preheated plasma was determined from the absolute Kα yield Y_k. We have assumed that the hot electrons have a Maxwellian velocity distribution function with a characteristic temperature T_h. We have calculated the efficiencies $g(T_h)$ of Kα emission[14] produced by fast electrons and the energy deposition $f(T_h,z)$[21] at different depths inside the aluminum layer as a function of the fast electron temperature. The deposited energy $E_{int}(z) = f(T_h,z).\phi$ and the Kα yield $Y_k = g(T_h).\phi$ (ϕ is the incident electron flux in J/cm²) can be combined to eliminate the incident electron flux. We have plotted in Fig.4 the ratio $g(T_h)/f(T_h,z)$ for different E_{int} (each E_{int} is directly related to an average Z_{eff} by the equation of state) as a function of the Kα yield and the hot electron temperature. We have chosen an average depth of 200 nm to agree with the experimental spatial deposition data. Due to the uncertainty in the surface of emission of the Kα lines, an upper limit of the measured Kα flux is $Y_k = 6 \cdot 10^{13}$ keV/sr/cm^2 (see Fig.4). The intersection of experimental data with theory give two possible hot electron temperatures, namely 0.9 keV and 10-30 keV. For each case, we can deduce the total energy in the fast electron component using the Kα efficiency (crosses in Fig.4).

For the 10-30 keV case, we found an electron flux of 200 J/cm^2 which represents 2% of the incident laser energy. These fast electrons can be generated by nonlinear processes in the laser plasma but they represent only a weak part of the absorbed energy. For the 0.9 keV case, the electron energy flux is 2 kJ/cm^2 which corresponds to 20% of the laser energy. In this case, the Kα production and the heating of the inner plasma are done principally by the electrons of the tail of the Maxwellian electron energy distribution. These electrons have a mean free path much longer than the electron temperature gradient scale length.

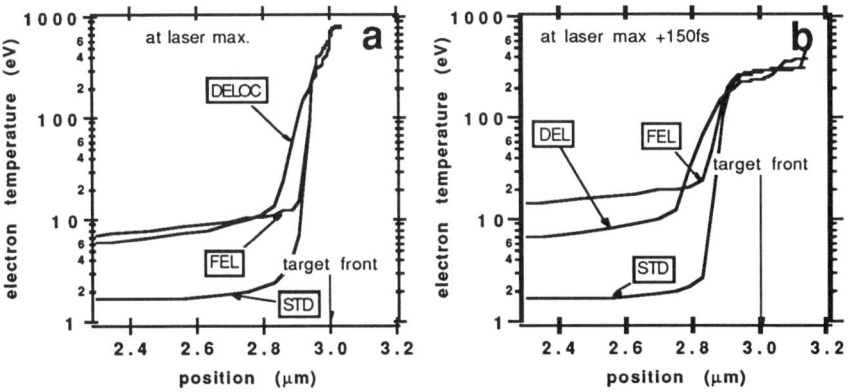

Fig 5 *Electron temperature profile given by the hydrodynamique simulation with fast electron transport (FEL) and the delocalized heat flux theory (DEL). The standard flux limited (0.1) Spitzer-Harm conduction is labelled STD.*

In order to discriminate between the two possible hot electron temperatures, we have used the 1D Lagrangien code FILM to simulate the femtosecond interaction. The Fresnel formulas for absorption in a solid were solved[10,22] at each time step to evaluate the laser absorption. The simulations were done for a solid aluminum target illuminated at normal incidence with a gaussian laser pulse of 100 fs FWHM and a laser intensity of 10^{17} W/cm^2. We have used the code with two different hypothesis for the control of the energy conduction. In the first case, we put 2% of the laser energy in the fast electron component with a temperature of 20 KeV at the maximum of the pulse[8]. In the second case, we removed the fast electron source and we used the delocalized heat flux theory[7] to test the hypothesis of target preheat by the quasi-ballistic electrons of the tail of the Maxwellian distribution function.

Results are given in Fig.5 for two different times during the laser interaction: at laser maximum (Fig.5a) and 150 fs after laser maximum (Fig.5b). The temperature profile obtained in both simulations show that the fast electron model (hereafter referred to as FEL) or the delocalized heat transport theory (hereafter referred to as DEL) give a similar electron temperature plateau inside the solid. The curve labelled STD is the result of a simulation where the heat flux is limited to 1/10 the value given by the standard Spitzer-Harm theory of heat conduction. The temperature found in the case of the DEL simulation is lower than in the case of the FEL simulation. The measured energy penetration depth is correctly reproduced in both simulations. It should be emphasized that the FEL simulation uses the experimental value (2%) for the hot electron deposition while the DEL simulation uses self consistently the calculated absorption which is only 7% as compared to the measured 20-30%. In both simulations the source of target preheat (and thus the Kα production) decreases sharply at the end of the laser pulse. Within the 10-30 keV fast electron hypothesis, the time duration of the electron source, which resides in non-linear processes of laser interaction, is comparable to the laser pulse duration[23]. In the delocalized heat transport hypothesis, the Kα production is quenched when the electron temperature drops at the end of the laser pulse due to the exponential variation of the number of electrons in the distribution tail with temperature. Owing to the transient nature of the production of energetic electrons and to their finite transit time over a deposition depth, Kα emission is not longer than the pulse duration.

In conclusion, femtosecond laser matter interaction generates intense Kα emission. The intensity of the Kα lines is about 5 TW/cm^2 in our experimental conditions. This emission can be associated to non local heat transport and/or to non-linear processes of electron production. If the non-linear hypothesis is retained, measurements give only 2% of the incident laser energy involved in the absorption process while in the delocalized transport hypothesis only 7% absorption is calculated. This is too low to explain the 20-30% measured absorption. Future work is needed to elucidate the rather large discrepancy found in the absorbed laser energy between the code and the experiment. Our analysis of x-ray emission suggest that a 100 fs X ray pulse of Kα lines can be produced when a 100 fs laser interacts with a metallic target. This is shorter than any other x-ray production by thermal mechanisms inside the plasma.

We would like to acknowledge F. Amiranoff for helphul discussions and a loan of programs for hot electron energy deposition calculations and F. Robelin (Quartz et Silice) for his help in the curved crystal design.

REFERENCES

1. R.W. Falcone and M.M. Murnane, in *Short Wavelength Coherent Radiation*, ed. D.T. Attwood and J. Bokor (American Institute of Physics, New-York, 1986), p.81.
2. O.R Wood II, *et.al.*, Appl. Phys. Lett. 53, 654 (1988).
3. M.M. Murnane, H.C. Kaypten, R.W. Falcone, Phys. Rev. Lett. 62, 155 (1989).
4. H.M. Milchberg, R.R. Freeman, S.C. Davey, R.M. More, Phys. Rev. Lett. 61, 2364 (1988).
5. Y.T. Lee and R.M. More , Phys Fluids 27, 1273 (1984).
6. J.A. Cobble, *et.al.*, Phys. Rev. A39, 454 (1989).
7. A. M. Mirza , G. Murtaza, Physica Scripta 41, 202 (1990) and references therein.
8. D. Shvarts, C. Jablon, I. Bernstein, J. Virmont P. Mora, Nucl. Fus. 19, 1471 (1979).
9. R.M. More, Adv. At. Mol. Phys. 21, 305 (1985).
10. H.M. Milchberg, R.R. Freeman, J. Opt. Soc. Am. B 6,1351 (1989).
11. J. Etchepare, G. Grillon, A. Orszag, IEEE Journal Quant. Elec. 19, 775 (1983).
12. J.P. Geindre, P. Audebert, J.C. Gauthier, R. Benattar, J.P. Chambaret, A. Mysyrowicz, A. Antonetti, Proceedings of the 21st Anomalous Absorption Conference, Banff, Canada (1991).
13. A. Zigler, P.G. Burkhalter, D.J. Nagel, M.D. Rosen, K. Boyer, G. Gibson, T.S. Luk, A. McPherson and C.K Rhodes, Appl. Phys. Lett. 59,534 (1991).
14. R.J. Harrach and R.E. Kidder, Phys. Rev. A23,887 (1981).
15. V.A. Boiko, S.A Pikuz and A. Ya. Faenov, J. Phys. B. 12, 1889 (1979).
15. R.W. Lee, B.L. Whitten and R.E. Strout, J. Quant. Spectrosc. Radiat. Transfer, 32,91(1985).
16. V. L. Jacobs, M. Blaha, Phys. Rev. A21, 525 (1980).
17. J.D. Hares, J.D. Kilkenny, M.H. Key and J.G. Lunney, Phys. Rev. Letters, 42,1216, (1979).
18. C. Chenais-Popovics, C. Fievet, J.P. Geindre, J.C. Gauthier, E Luc-Koenig, J. F. Wyart, H. Pepin and M. Chaker, Phys. Rev. A40, 3194, (1989).
19. R.M. More, K.H. Warren , D.A. Young, and G.B. Zimmerman Phys. Fluids.31, 3059 (1988).
20. I.V. Spencer, Phys. Rev. 98, 1597 (1955).
21. M. Chaker, J.C. Kieffer, J.P. Matte, H. Pepin, P. Audebert, P. Maine, D. Strickland, P. Bado, G. Mourou, Phys. of Fluids 3,167 (1991)
23. N.H. Burnett, G.D. Enright, A. Avery, A. Loen and J.C. Kieffer, Phys. Rev. A29, 2294 (1984).

RECENT DEVELOPMENTS IN THE SUPER TRANSITION ARRAY MODEL FOR SPECTRAL SIMULATION OF LTE PLASMAS

A. Bar-Shalom and J. Oreg
Nuclear Research Center Negev, P.O.Box 9001, Beer-Sheva 84190, Israel

W. H. Goldstein
Lawrence Livermore National Laboratory, Livermore, California 94550

ABSTRACT

Recently developed sub-picosecond pulse lasers have been used to create hot, near solid density plasmas. Since these plasmas are nearly in local thermodynamic equilibrium (LTE), their emission spectra involve a huge number of populated configurations. A typical spectrum is a combination of many unresolved clusters of emission, each containing an immense number of overlapping, unresolvable bound-bound and bound-free transitions. Under LTE, or near LTE conditions, traditional detailed configuration or detailed term spectroscopic models are not capable of handling the vast number of transitions involved. The average atom (AA) model, on the other hand, accounts for all relevant transitions, but in an oversimplified fashion that ignores all spectral structure. The Super Transition Array (STA) model, which we have developed in recent years, combines the simplicity and comprehensiveness of the AA model with the accuracy of detailed term accounting. The resolvable structure of spectral clusters is revealed by successively increasing the number of distinct STA's, until convergence is attained. The limit of this procedure is a detailed unresolved transition array (UTA) spectrum, with a term-broadened line for each accessible configuration-to-configuration transition, weighted by the relevant Boltzman population. In practice, we have found that this UTA spectrum is actually obtained using only a few thousand to tens of thousands of STA's (as opposed, typically, to billions of UTAs). The central result of STA theory is a set of formulas for the moments (total intensity, average transition energy, variance) of an STA. In calculating the moments, we use detailed relativistic first order quantum transition energies and probabilities. The energy appearing in the Boltzman factor associated with each level in a superconfiguration is the zero order result corrected by a superconfiguration averaged first order correction. As the number of configurations in a superconfiguration is successively decreased, this becomes equivalent to using exact first order configuration average energies. In addition, orbital relaxation can be accounted for by recalculating orbitals and energies for each STA in a potential optimized for the particular set of configurations. Examples and application to recent measurements are presented

INTRODUCTION

Hot, dense plasmas occur in physical systems as diverse as astrophysical objects and laboratory fusion experiments. The recent development of sub-picosecond pulse lasers has opened new research possibilities for the study of near solid density plasmas. A central diagnostic for laser-produced plasmas involves the interpretation of their emission spectra. However, under high density, near-LTE conditions, a huge number of ionic configurations are populated; a typical number for heavy atoms is 10^{15}. Traditional spectroscopic approaches are difficult to apply in such cases. A typical emission spectrum of a sub-picosecond laser-produced plasma, obtained by Zigler and coworkers[1] using 650 fs KrF* laser pulses and a BaF$_2$ target, is shown in Fig. 1. This kind of spectrum is comprised of many unresolved clusters, each containing an immense number of overlapping bound-bound and bound-free transitions.

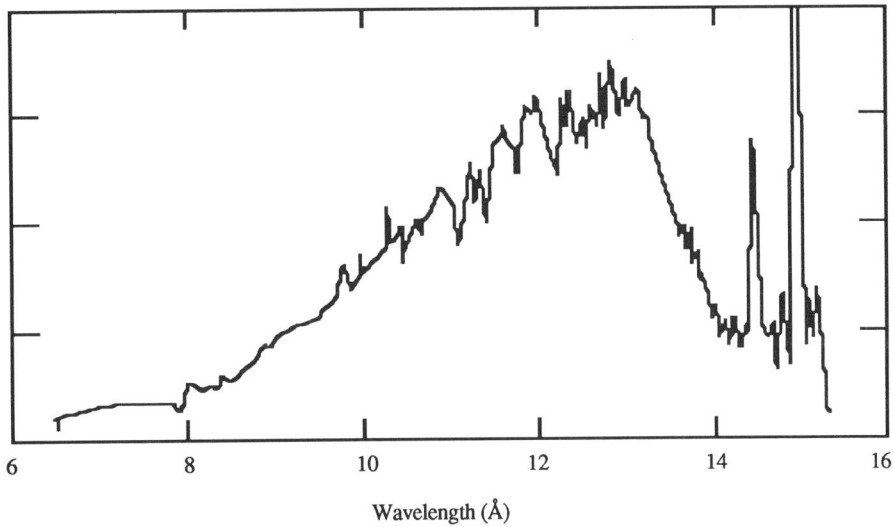

Fig. 1 BaF$_2$ spectrum obtained in ultra-short pulse laser-plasma experiment reported in Ref. 1.

The spectrum is a sum over all transitions from all populated levels:

$$I(\nu) = \Sigma_{l,l'} N_l f_{l,l'} P(\nu - E_{l,l'}) \tag{1}$$

where N_l is the initial level population $f_{l,l'}$ is the transition probability from level l to level l' and P is a normalized line shape function which is in general a Voigt convolution of the Doppler and pressure broadened profiles. In LTE, the level populations N_l are determined by the Saha-Boltzmann equations. All the additional relevant information needed to describe the spectrum – the transition energies, $E_{l,l'}$ and transition probabilities $f_{l,l'}$ – may be obtained from standard atomic structure models. The essential problem in the simulation of these spectra is accounting for the enormous number of populated levels.

Models that account explicitly for each and every transition line, such as the OPAL code of Rogers, Iglesias and Wilson,[2] are not suited for treating the immense number of lines that comprise the spectra of moderately stripped heavy atoms such as barium. Therefore models were developed that describe a manifold of unresolved transitions by the moments of its spectral distribution: total intensity,

$$I = \int I(\nu) d\nu = \Sigma_{l,l'} N_l f_{l,l'} \tag{2}$$

average energy,

$$E = [\Sigma_{l,l'} N_l f_{l,l'} E_{l,l'}] / I \tag{3}$$

and variance,

$$\Delta E^2 = [\Sigma_{l,l'} N_l f_{l,l'}(E_{l,l'} - E)^2]/I \quad , \tag{4}$$

where the summations are over all the transitions belonging to the manifold.

The expressions in Eqs. (3), (4) and (5) still involve explicit summation over all transitions. However, for certain limiting choices of transition manifold, algebraic summations have been obtained. One such limit is the configuration-configuration manifold, or Unresolved Transition Array (UTA), developed by Bauche et al.[3] In this model each configuration-to-configuration manifold of transitions is represented by a Gaussian defined by these moments. This model avoids dealing with individual levels and is most useful when only a relatively small number of configurations are populated. For heavy ions in LTE, though, the huge number of relevant UTAs in each one-electron transition array, makes it generally impractical.

Another limit was considered by Stein, Shalitin and Ron.[4] With some simplifying assumptions they obtained the spectral moments by algebraic summation for the case where the manifold includes all possible contributions to a one-electron transition. In this model, all contributions are accounted for, but each such array is described by a single Gaussian without internal structure. This method is very efficient but does not reveal the details of the spectrum.

SUPER TRANSITION ARRAYS

The Super Transition Array (STA) approach[5] is to split each one-electron transition array into as many partial arrays, or sub-arrays, as needed to reveal the detailed structure of the spectrum. We have derived analytical expressions for the moments of each sub-array and represent it by a separate Gaussian. Thus the details of the spectrum is revealed gradually as we refine the splitting and increase the number of sub-arrays.

In order to specify the sub-arrays, we define supershells and superconfigurations. A supershell σ, is a collection of adjacent atomic sub-shells $s \in \sigma$, and a superconfiguration is specified when the occupation numbers, q_σ, of each supershell are given. The superconfiguration is simply the collection of all configurations resulting from all the possible distributions of the q_σ electrons in the shells $s \in \sigma$. It is therefore a collection of many near-lying configurations. The STA is now defined as the array of transitions between two superconfigurations. The number of individual STAs that are used to describe a one-electron transition determines the accuracy with which the spectrum is simulated and the degree to which its UTA structure is resolved. In the limit that each STA reduces to a single UTA, we end up with the detailed UTA spectrum. However, we have found that the spectrum converges to the UTA specrum after constructing a relatively small number of STA's.

THE STA MOMENTS AS GENERALIZED PARTITION FUNCTIONS

The derivation of algebraic formulas for the STA moments is carried out in two steps. First we express the moments in terms of general partition functions which have a simple standard form, but still include explicit summations over detailed levels. In the second step, described in the next section, these summations are performed using recursion formulas.

In order to obtain the moments of an STA in the form of general partition functions we first express them in terms of configuration average quantities:

$$I_A^{\alpha\beta} = \sum_{C \in A} N_C f_{CC'} \qquad C' = C - \alpha + \beta$$

$$E_A^{\alpha\beta} = \left[\sum_{C \in A} N_C f_{CC'} E_{CC'} \right] / I_A^{\alpha\beta} \qquad (5)$$

$$\left(\Delta E_A^{\alpha\beta}\right)^2 = \sum_{C \in A} N_C f_{CC'} \left[\left(E_{CC'} - E_A^{\alpha\beta}\right)^2 + \Delta(C,C) \right] / I_A^{\alpha\beta}$$

The summations here are on configurations C' contained in superconfiguration A, and not on detailed levels. But the quantities $f_{CC'}$, $E_{CC'}$ and $\Delta(C,C')$ are defined[3,5] so that the moments are exact, and amount to detailed summations over levels with no approximations. The configuration C' is uniquely determined from C by the specific transition, $\alpha \Rightarrow \beta$, under consideration.

The configuration averaged quantities for

$$C = \prod_s (nl)_s^{q_s} \qquad (6)$$

and C' are given in terms of the occupations numbers q_s, as[5]

$$N_C = e^{-\frac{\Delta E_A^1}{kT}} \prod_s \binom{g_s}{q_s} X_s^{q_s}, \qquad X_s = e^{-\frac{\varepsilon_s - \mu}{kT}}$$

$$f_{CC'} \propto q_\alpha (g_\beta - q_\beta)$$

$$E_{CC'} = D_0 + \sum_s (q_s - \delta_{s\alpha}) D_s \qquad (7)$$

$$\Delta(C,C) = \sum_s (q_s - \delta_{s\alpha})(g_s - \delta_{s\alpha} - \delta_{s\beta}) \Delta_s$$

Here $(nl)_s$ designates a non-relativistic or relativistic atomic sub-shell, and g_s and g_c are the statistical weights of the shell s and of the configuration C, respectively.

The configuration average energy entering the Boltzman factor for configuration C, N_C, has been factored into a zeroeth order term that is a sum of single orbital energies, ε_s, and a first order correction ΔE_A^1, which is an average value of this correction for the superconfiguration A. Since this contribution is common to every configuration, it factors from the sums in Eq. (5). The orbital energies, ε_s, and the quantities D_s and Δ_s are independent of the occupation numbers and are also common to all the configurations in the STA.

The generalized partition function form of the moments is obtained now by subtitution in Eq. (5) of the above quantities and simple manipulations of the binomial coefficients. We will demonstrate the algebra for the STA total intensity only. The substitution gives

$$I_A^{\alpha\beta} = \sum_{C \in A} N_C f_{CC'} \propto \sum_{C \in A} q_\alpha (g_\beta - q_\beta) \prod_{s \in C} \binom{g_s}{q_s} X_s^{q_s} \tag{8}$$

In this expression and the summation over all configurations is in fact a summation over all the occupation numbers of the configurations which belong to the inital superconfiguration. Now, with zeroeth order configuration energies, the standard form of a partition function of a superconfiguration is

$$U_Q(A) \equiv \sum_{C \in A} g_C e^{-\frac{E_C^{(0)}}{kT}} = \sum_{C \in A} \prod_{s \in C} \binom{g_s}{q_s} X_s^{q_s}, \tag{9}$$

and we see that in order to identify Eq. (8) with (9), we must eliminate the multiplicative factor $q(g-q)$ under the summation in (8). This is achieved using the standard binomial relations

$$q\binom{g}{q} = g\binom{g-1}{q-1}, \quad (g-q)\binom{g}{q} = g\binom{g-1}{q}, \tag{10}$$

to replace each q dependent factor by a g dependent factor that is common to all the terms in the sum. In this way we obtain

$$I_A^{\alpha\beta} \propto \sum_{C \in A} \prod_{s \in C} \binom{g_s^{\alpha\beta}}{q_s^\alpha} X_s^{q_s^\alpha} = U_{Q-1}^{\alpha\beta}(A) \tag{11}$$

This result has the form of a partition function, with modified statistical weights and occupation numbers,

$$g_s^{\alpha\beta} = g_s - \delta_{s\alpha} - \delta_{s\beta}, \quad q_s^\alpha = q_s - \delta_{s\alpha}. \tag{12}$$

The formal partition function, however, still includes the summations over all configurations of the initial superconfiguration. This complexity is addressed in the next section.

RECURSION FORMULAS FOR THE PARTITION FUNCTIONS

Using the generating function,

$$F(z) \equiv \sum_{Q_\sigma} z^{Q_\sigma} U_{Q_\sigma} = \prod_{s \in \sigma} (1 + zX_s)^{g_s}, \quad (13)$$

we have obtained[5] the following recursion formula for the supershell σ:

$$U_{Q_\sigma}(g) = \frac{1}{Q_\sigma} \sum_{n=1}^{Q_\sigma} \chi_n U_{Q_\sigma - n}(g), \quad \chi_n = -\sum_{s \in \sigma} g_s (-X_s)^n \quad (14)$$

The total partition function for the superconfiguration is then

$$U_Q = \prod_\sigma U_{Q_\sigma} \quad (15)$$

It is clear from Eqs. (14) and (15) that working recursively from $U_0=1$, the summation in Eq.(14) contains only a few tens of terms, rather than the astronomical number of terms in the original expressions for the moments. Once the approximation of zeroeth order energies is made for the Boltzman energies, this result is a mathematical identity with no approximations.

Similar recursion formulas were obtained starting from the full supershell. In addition, we obtained relations between partition functions with general statistical weights that are used to speed up the calculations of all the STA moments.

RESULTS

The efficiency and accuracy of the STA model is demonstrated in Fig. 2, for the $2p_{3/2}$-$3d_{5/2}$ transition array of iron, at an electron temperature of 200 eV and solid density. Under these plasma conditions, the number of bound orbitals is around 14 and these give rise to a modest 212,049 UTAs, a number small enough to be calculated one-by-one. It is seen that our model converges to the exact UTA result with only 673 STAs. On the other hand taking only 69 STAs still hides the main features of the structure. Clearly the Average Atom (AA) model which utilizes a single Gaussian for the whole one-electron transition array fails to describe the correct structure.

In Fig. 3 an STA model for the ultra-short pulse Ba laser-plasma spectrum is shown. The solid trace is the sum of the spectra obtained for each temperature, weighted by the fractions in parentheses. The weights were chosen to give a reasonably good fit to the data shown in Fig. 1. Though this is neither a precise procedure nor a unique result, it conveys the essential point that the STA model can be used to characterize the emission by a distribution of temperatures between 200 and 350 eV, and that the peak emission corresponds to a temperature around 225 eV.

The remainder of our examples are simulations of recent transmission measurements carried out by Foster et al.[6] Point-projection spectroscopy was used to measure the transmission of 1-2 KeV X-rays through 3000 Å of germanium. By means including hydrodynamic simulation and radiography, the sample conditions in were determined to

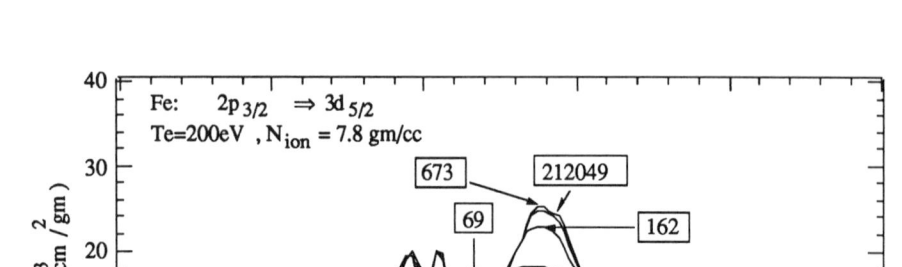

Fig. 2. Spectrum convergence with increasing number of STAs

be 76 eV and .05 g/cc. In the calculations below, all bound-bound and bound-free[7] transitions are calculated using the STA model, while free-free transitions are handled using Kramers rule, and the scattering contribution is calculated from Sampson's Compton scattering results.[8] Plasma effects are included using an ion-sphere ansatz and the Thomas-Fermi model for the free electrons. The bound orbitals are obtained by first order solution to the 'many-electron' Dirac equation consistent with the free electron potential. The effect of collisional broadening is included by assuming a common Voight line shape, obtained by convolving a Doppler profile and collisional Lorentzian for each line within an STA. We convolve each STA Gaussian with this Voight line shape, to obtain a Voight function for each STA.

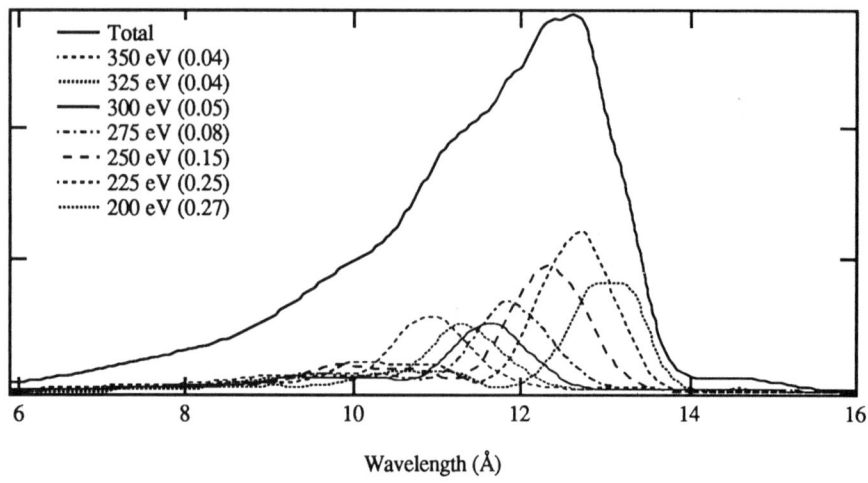

Fig. 3. STA model for Ba spectrum shown in Fig. 1. The solid trace is the sum of the spectra obtained for each temperature, weighted by the fractions in parentheses.

In Fig. 4 we present an STA model for the germanium transmission spectrum between 1100 and 1700 eV. The prominent features are the 2p-3d transition array at ~1300 eV, the 2s-3p at ~1380 eV, and the 2-n, n≥3, grouped above 1400 eV. This figure compares calculations including (solid trace) and neglecting the effects of term structure (UTA widths and shifts). It is interesting to note, in addition to the difference in structure, a small shift in the mean energies of several of the features, attributable to the UTA shift.

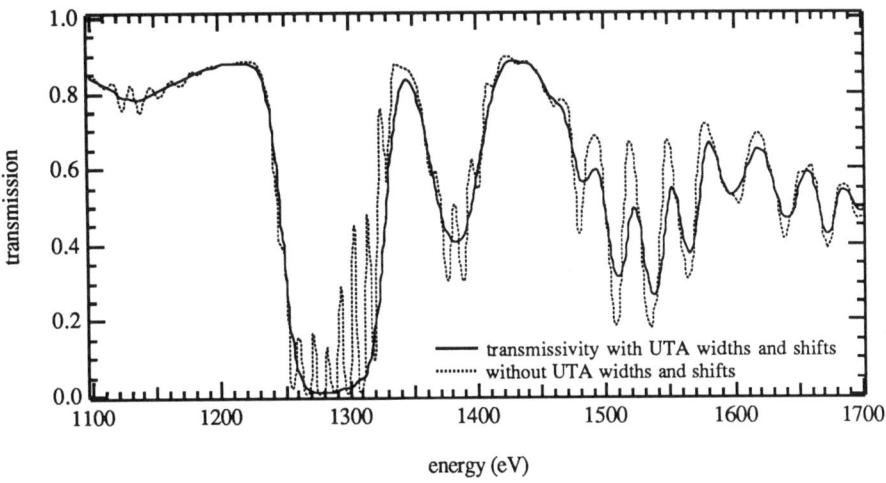

Fig. 4. STA models for transmissivity of 3000 Å of Ge at 76 eV and .05 g/cc.

In Fig. 5 we compare the model in Fig. 4, including term structure,, where a different, optimized parametric potential was used for the most important superconfigurations, with a model based on a single potential, optimized for the most likely superconfiguration. Two effects can be observed. First, the transition arrays are

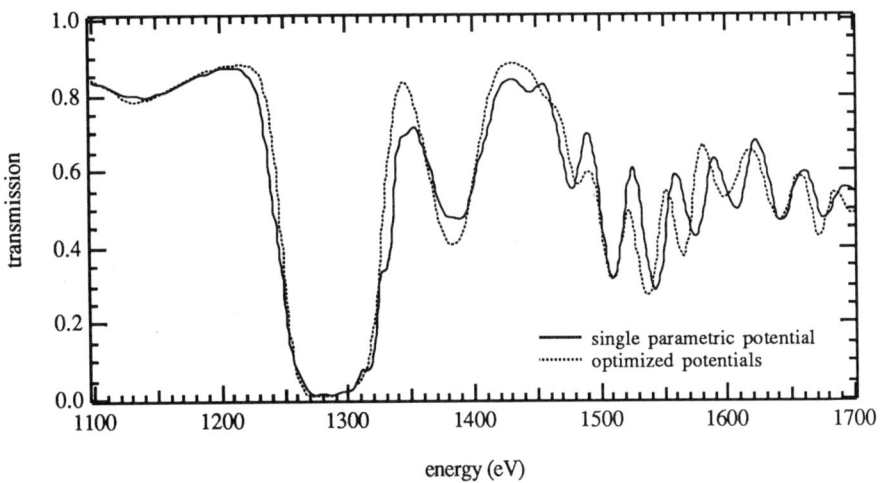

Fig. 5. Comparison of models using many optimized potentials and a single potential

broadened in the latter case, as seen most clearly in the 2p-3d feature. Second, the mean energies of transition arrays are shifted to higher energies, the more so at higher frequencies. Both of these effects are explained by the fact that the energy of a configuration is up-shifted when using orbitals which optimize the energy of another configuration, and this shift becomes greater as the energy distance between the configurations is increased.

Finally, Fig. 6 demonstrates how the calculation converges as more and smaller superconfigurations are included. The Average Atom calculation uses a single superconfiguration, i.e., one STA for each one-electron transition, while the 47,766 STA calculation is based on a 154 superconfigurations. By requiring 1% accuracy in both the group Rosseland and Planck mean free paths in the spectral range 1200-1700 eV, the 107,605 STA simulation was obtained (shown also as the solid trace in Fig. 4). Roughly 400 superconfigurations were generated in this latter case.

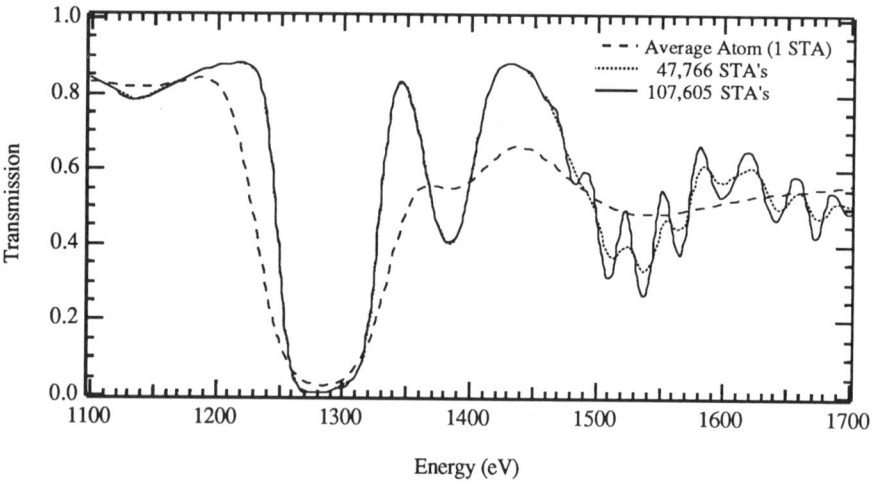

Fig. 6. Convergence of the STA simulation for Ge transmission, with increasing number of STA's

SUMMARY

We have presented the theoretical basics of the STA model for simulating bound-bound and bound-free transitions in LTE plasmas. Examples of the method have been given, as well as results relevant to current experimental programs, including ultra-short pulse laser-produced plasmas, and point-projection transmission measurements.

REFERENCES

1. A. Zigler, P. G. Burhalter, D.J. Nagel, K. Boyer, T.S. Luk, A. McPherson, J. C. Solem and C.K. Rhodes, Appl. Phys. Lett. 59, 777 (1991); and A. Zigler, P. G. Burhalter, D.J. Nagel, W.H. Goldstein, A. Bar-Shalom, T.S. Luk, A. McPherson and C.K. Rhodes, to appear in the proceedings of the International Workshop on Radiative Properties of Hot Dense Matter, Oct. 22-26, 1990, Sarasota, Florida.
2. F. J. Rogers, C. A. Iglesias and B. G. Wilson, Astrophys. J. 322, L45 (1987).
3. J. Bauche, C. Bauche-Arnoult, and M. Klapisch, Phys. Rev. A20, 2424 (1979).

4. J. Stein, D. Shalitin and A. Ron, Phys. Rev. A35, 280 (1987).
5. A. Bar-Shalom, J. Oreg, W. H. Goldstein, D. Shvarts and A. Zigler, Phys. Rev. A40, 3183 (1989)
6. J. M. Foster, D. J. Hoarty, C. C. Smith, P. A. Rosen, S. J. Davidson, S. J. Rose, T. S. Perry and F. J. D. Serduke, "L-shell Absorption Spectrum of an Open M-shell Germanium Plasma: Comparison of Experimental Data With a Detailed Configuration Accounting Calculation," submitted to Phys. Rev. Lett.
7. A. Bar-Shalom, J. Oreg and W. H. Goldstein, "Calculation of Emission and Absorption Spectra of LTE Plasma by the STA Method," to appear in the proceedings of the International Workshop on Radiative Properties of Hot Dense Matter, Oct. 22-26, 1990, Sarasota, Florida.
8. D. H. Sampson, Astrophys. J. 129, 734 (1959).

MEASUREMENTS AND MODELS OF THE OPACITY OF HOT, DENSE PLASMA

P.T. Springer, T.S. Perry, D.F. Fields, W.H. Goldstein,
B.G. Wilson, and R.E. Stewart
Lawrence Livermore National Laboratory, Livermore, California 94550

ABSTRACT

In order to test and improve opacity models, we are performing experiments to accurately characterize and measure the opacity of high-Z plasmas. In the experiments the NOVA laser at Lawrence Livermore National Laboratory is used to indirectly heat tamped thin foil targets containing mid to high-Z ions of interest which are co-mixed with low-Z ions for spectroscopic determinations of plasma conditions. One of the NOVA beams is used to create a source of x-rays for backlighting the plasma face-on, while another beam is used to perform a radiograph of the expanding sample edge-on. X-rays that pass through and around the plasma are simultaneously measured on high resolution crystal spectrometers, and the opacity is measured by comparing the direct and attenuated x-ray spectra. Plasma density is determined from the spatial extent of the expanded plasma, and plasma temperature is inferred from the ionization distribution of the low-Z ions doped in the plasma. Predictions of the high-Z opacity are then compared with data using measured conditions as input. These are the first high temperature opacity measurements in which sample conditions are simultaneously determined. We present measurements and predictions for a spectroscopic experiment using a comixture of aluminum and niobium. We discuss plans for future improvements, and for spectroscopic measurements in the spectral region determining Rosseland mean opacities.

INTRODUCTION

Knowledge of the x-ray absorption properties of hot, ionized plasma in thermodynamic equilibrium is important in fields ranging from stellar astrophysics to inertial confinement fusion. Many plasma properties are affected by opacity including coupling and the flow of radiation through plasma, plasma dynamics, and plasma emission. Opacity models are complex and require knowledge of atomic energy levels, oscillator strengths, atomic level populations, line-shapes, and ion-plasma interactions. For most plasmas of interest opacity calculations require simplifying assumptions due to the enormous number of configurations contributing in a fully detailed calculation, and to the insoluble nature of the many-body problem. Due to the complexity and widespread applicability of opacity calculations, it is important to benchmark them with laboratory measurements.

MEASUREMENT TECHNIQUES

In order to test and improve opacity models, we are performing laboratory experiments to measure the opacity of complex, high-Z plasmas.[1] Because opacity models depend critically upon both the plasma temperature and density, it is important to characterize the plasma conditions in addition to measuring its opacity. The experimental setup is shown in Figure 1. The NOVA laser, at Lawrence Livermore National Laboratory, is used to indirectly heat tamped targets containing mid to high-Z ions of interest which are co-mixed with low-Z ions for spectroscopic determinations of plasma conditions. As the sample ionizes and explodes, the tamping material maintains a uniform spatial density profile in the opacity sample One of the NOVA beams fired 2 ns after the main heating beams is used to create a 200 ps source of x-rays for backlighting the plasma face-on, and measure its absorption in two separate spectral regions. The plasma density is measured at the same time that its opacity is measured by firing another Nova beam producing a point source of x-rays to radiograph the expanding sample edge-on. X-rays that pass through and around the plasma are simultaneously measured on a high resolution crystal spectrometers using PET crystals, and the opacity is measured by comparing the direct and attenuated x-ray spectra.

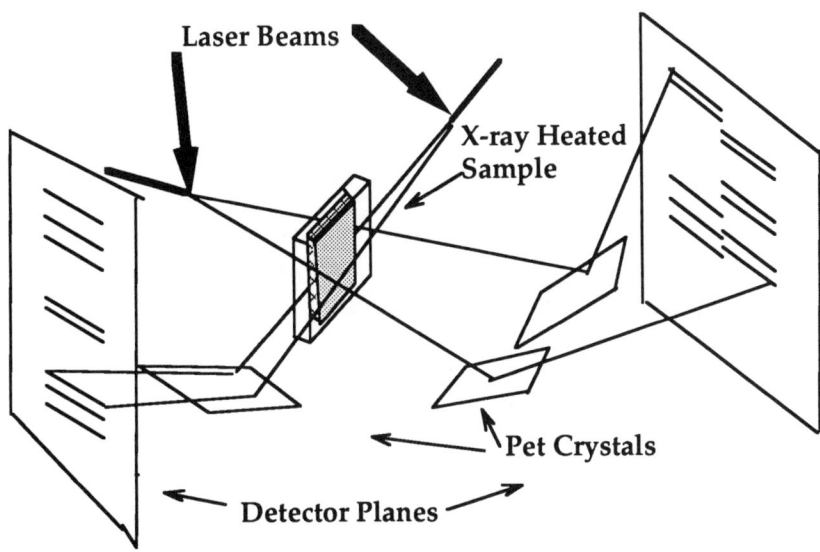

Fig. 1. Experimental setup.

The thin foil sample consisted of 222 µg/cm^2 of niobium comixed with 35 µg/cm^2 of aluminum. The thicknesses of the aluminum and niobium were measured using x-ray florescence with an accuracy of order 5 percent. Figure 2 shows the face-on data showing absorption spectra of aluminum in the 1500 - 2000 eV region on one PET crystal, and the absorption of niobium in the 2000-3000 eV region on another. Figure 3 shows the absorption data for the side-on view on this shot. The spectral resolution E/δE for this experiment was of order 2000. Data was also obtained by backlighting at 2.5 nanoseconds, and the effect of the plasma expansion was clearly seen in the radiograph, as well as in the ionization distribution of the aluminum. The film density versus position was measured using a micro-densitometer. An x-ray film calibration was applied to convert from density to x-ray exposure. The curvature of the lines due to the spectrometer geometry were removed. Emission and scattered backgrounds limit the apparent transmission in spectral regions were the plasma is optically thick. These backgrounds which are evident in regions off the crystal and outside of the image of the diagnostic apertures were estimated and subtracted from the data. The plasma transmission versus x-ray energy was then obtained by dividing the attenuated and direct backlighter spectrum. The energy calibration is performed with sulfur and silicon emission lines observed in the spectrum.

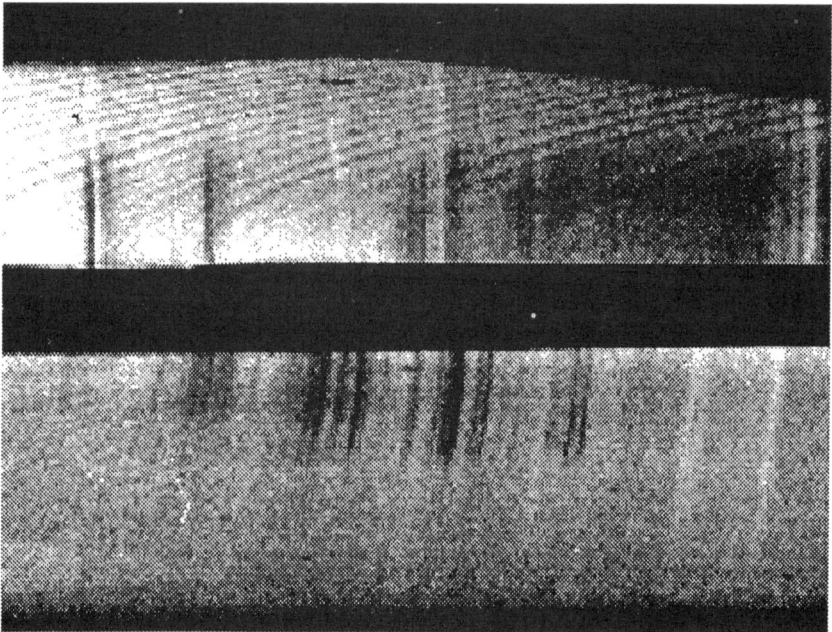

Fig. 2. Face-on point-projection opacity data for aluminum niobium plasma.

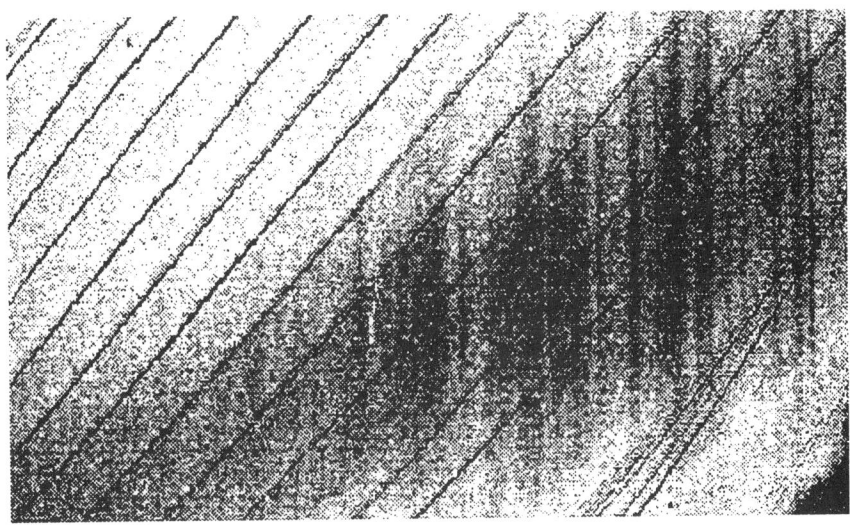

Fig. 3. Side-on radiography for aluminum niobium plasma. The diagonal lines were caused by crystal defects, and were filtered for the analysis.

PLASMA CHARACTERIZATION ANALYSIS

The plasma density was determined to be 0.0257 +/- 0.0005 g/cc using the measured width of the expanded plasma observed in the radiograph This width includes corrections for the finite size of the backlighting plasma and for initial bowing observed in the target. Because of careful characterization of the initial target composition and thickness the density uncertainty is limited only by these resolution corrections. Because of its low opacity, the plastic tamper is not visible in the radiographs, and there is no difference between the width inferred from the aluminum line absorption and the width inferred from the niobium bound-free absorption. In addition to giving the plasma density, the use of the radiographic technique also increases the dynamic range of the absorption spectroscopy technique due to the increase in sample thickness viewed edge-on, allowing measurements of weakly absorbing features. This is important for accurately determining the degree to which plasmas transmit, which will be important in obtaining spectroscopic measurements of Rosseland mean opacities.

Having determined plasma density, the temperature is inferred from the ionization distribution of the aluminum in the plasma. Note that the data shown in fig. 2 shows a slight gradient in aluminum ionization between the center and edge of sample. Figure 4 shows the measured aluminum absorption spectrum showing the presence of absorption from carbon-like through lithium-like aluminum, and the absence of helium-like features.

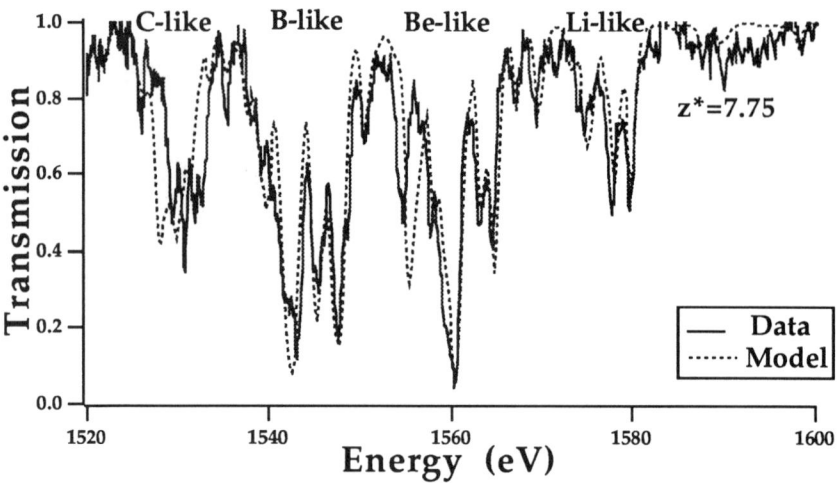

Fig. 4. Aluminum data and model.

Also shown in Figure 4 is a theoretical opacity calculation[2] based on highly accurate atomic structure data. This calculation, however, does not account for the lowering of the continuum due to plasma screening, and tends to underestimate the ionization for a given temperature. The abundances of the carbon-like through lithium-like aluminum were obtained from the calculation which matches the data closely, giving an average ionization Z* of the aluminum of 7.75. The opacity models MCUTA and STA, described below were then used to predict the temperature at which the aluminum ionization matches that of the spectroscopically accurate calculation, with the density constrained by the radiographic data. The temperature derived in the manner was 47 +/- 3 eV. The 20 percent uncertainty in plasma density corresponds to only a 1 eV uncertainty in plasma temperature. The 3 eV uncertainty in the temperature is estimated from the sensitivity of plasma ionization to a number of different opacity models without regard to which model is likely to be most accurate.

OPACITY DATA AND THEORY

With the plasma temperature determined from the aluminum ionization, and density determined from the radiography, predictions of the opacity of niobium are then compared with data. Two different opacity calculations are used to compare with data. The first model, MCUTA, uses a Monte Carlo Unresolved Transition Array accounting model, which employs a Monte Carlo random walk between configurations based on atomic transition probabilities. This algorithm is run until the resulting opacity spectrum converges. The second model, STA, uses the Super Transition Array theory[3,4,5] in which the total transition array of a single-electron transition, including all contributing configurations, is described by a small number of Super Transition Arrays, STA's, whose transition energies, intensities and variances are evaluated in optimized potentials. The number of distinct STA's in the calculation is increased until the spectrum converges. Both models use J-J coupling and an ion sphere ansatz to describe plasma effects.

Figure 5 shows the transmission of the aluminum niobium plasma deduced from the data. Also shown are the results of the two opacity calculations. Both calculations have been corrected for an instrumental resolution of approximately 1 eV. In general, the agreement between theory and experiment is excellent for both theoretical calculations, especially considering that these models are not designed for spectroscopic accuracy. However, there are some differences in structure which are evident in Figure 5. Figure 6 shows a comparison between theory and experiment in the region of the 2p-3d transition arrays. The lines have been shifted by approximately 15 eV in energy to overlay the data. The agreement between data and theory is excellent in both intensity, width, and admixture of components from differing ion stages. It is important to include orbital relaxation effects to match the observed transition widths.

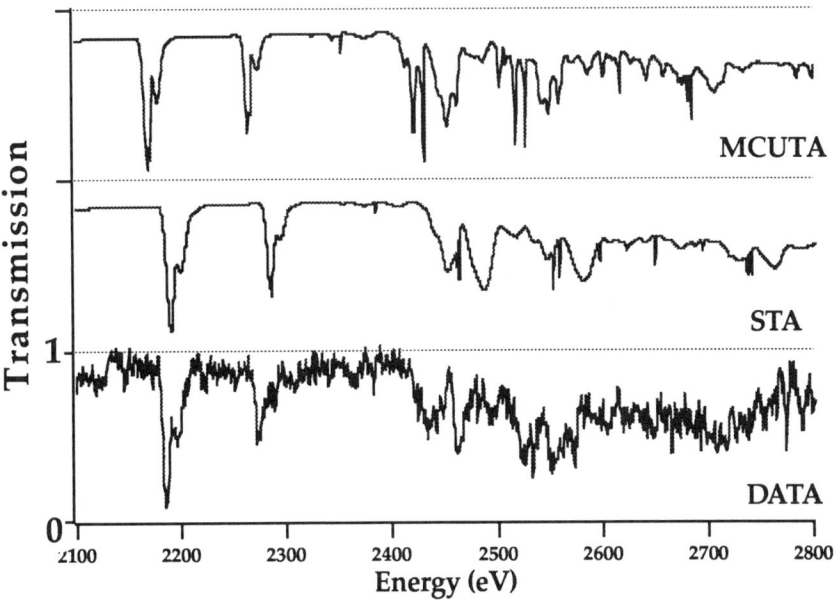

Fig. 5. Niobium data and models.

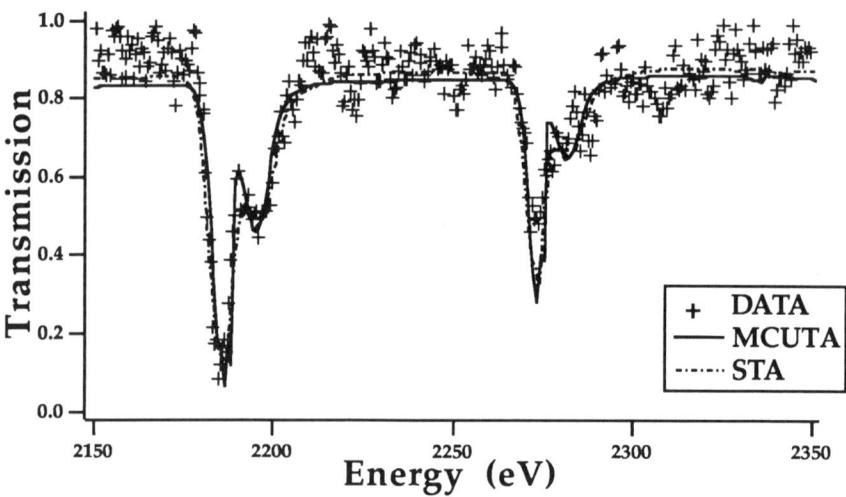

Fig. 6. Niobium 2p-3d data and models.

FUTURE EXPERIMENTS

We are extending measurement techniques into the soft x-ray and XUV regions, with the goal of performing opacity measurements in the region where the Rosseland mean opacity is determined. Measurements in this region can directly impact radiation flow calculations. We have developed gated crystal spectrometers, and new imaging techniques to reduce and quantify emission backgrounds. We plan to record and analyze sample emission to as a diagnostic of plasma conditions, and to verify that LTE conditions are established via an application of Kirchoff's laws, relating plasma emission and absorption. To achieve better plasma characterization, and to test light-element opacity models we wish to utilize temperature diagnostics that are less model dependent than the aluminum ionization. It is not satisfactory to depend upon the plasma ionization to derive temperature since many of the plasma ion interactions affecting opacity also affect the temperature determination. In the future, the plasma temperature will be determined with line-ratio techniques in helium-like and lithium-like sodium and fluorine ions where the oscillator strengths, energies, and number of relevant configurations can be more simply and accurately calculated. In addition, an independent temperature measurement using free-bound continuum emission will be attempted with the completion of XUV spectrometers currently under construction.

SUMMARY

We have described experiments to accurately characterize and measure the opacity of high-Z plasmas using point-projection spectroscopy. Plasma density is determined from the spatial extent of the expanded plasma, and plasma temperature is inferred from the ionization distribution of the low-Z ions doped in the plasma. These are the first high temperature opacity measurements in which sample conditions are simultaneously determined. Predictions of the high-Z opacity are then compared with data using measured conditions as input. We present measurements and predictions for a spectroscopic experiment using a comixture of aluminum and niobium. We discuss plans for future improvements, and for spectroscopic measurements in the spectral region determining Rosseland mean opacities.

ACKNOWLEDGEMENTS

We would like to thank C. Bruns and D. Nelson for helping assemble the experiment, J. Ticehurst for her help with the analysis, R. Wallace for his assistance in target preparation, J. Abdallah Jr. and R.E.H. Clark for their calculations, and the NOVA crew for operating the laser. This work was performed under the auspices of the U.S. Department of Energy by the Lawrence Livermore National Laboratory under Contract No. W-7405-ENG-48.

REFERENCES

1) T.S. Perry, S.J. Davidson, F.J.D. Serduke, D.R. Bach, C.C. Smith, J.M. Foster, R.J. Doyas, R.E. Stewart, J.D. Kilkenny, and R.W. Lee, "Opacity Measurements in Hot Dense Plasma," submitted to Phys. Rev. Letts.
2) J. Abdallah Jr. and R.E.H. Clark, J. Appl. Phys. 69, 23 (1991)
3) A. Bar-Shalom, J. Oreg, W.H. Goldstein, D. Shvarts, and A. Zigler, Phys. Rev. A40,6 3183 (Sept 1989)
4) A. Bar-Shalom, J. Oreg, and W.H. Goldstein, "Calculation of Emission and Absorption Spectra of LTE Plasma by the STA Method," to appear in the proceedings of the International Workshop on Radiative Properties of Hot Dense Matter, Oct. 22-26, 1990, Sarasota Florida
5) A. Bar-Shalom, J. Oreg and W.H. Goldstein, "Recent Developments in the Super Transition Array Model for Spectral Simulations of LTE Plasma, in these proceedings

THE INNER-SHELL 3d-4f TRANSITIONS IN THE SPECTRA
OF HIGHLY IONIZED HEAVY ELEMENTS

P. Mandelbaum*
E.O. Hulburt Center For Space Research
Naval Research Laboratory
Washington,DC 20375-5000

ABSTRACT

Many X-ray spectra emitted by laser-irradiated high-Z elements are dominated by the intense pseudocontinuum originating from 3d-4f inner-shell resonance transitions. In this paper, some of the particular properties of the 3d-4f inner-shell excited levels are presented and the implications on the interpretation of the observed spectra are discussed.

INTRODUCTION

Many X-ray spectra of highly ionized elements emitted from laser-produced plasma are characterized by the dominant 3d-4f pseudocontinuum emitted by the ionization states neighbouring the Ni-like ion. First report of this kind of spectrum has been given to our best knowledge by Burkhalter et al.[1] for a Gd spectrum . Since this pioneering experimental work, many similar spectra have been obtained for almost all the elements up to highly ionized uranium. Figure 1 shows a typical spectrum from laser-irradiated lanthanum obtained at the Soreq Nuclear Center. The 3d-4f resonance transitions dominate the spectrum. These transitions can be divided into two subgroups. The first group includes the outer-shell 3d-4f transitions in ionization states higher than the Ni-like ionisation state. These are transitions of the $3d^n$-$3d^{n-1}4f$ kind (n=1-10) . These transitions appear on the shorter wavelength side of the $3d^{10}$-$3d^9 4f$ Ni-like transitions. The second group includes the 3d-4f inner-shell transitions in the Cu-like and lower ionization states. These are transitions of the $3d^{10}4l..4l'$-$3d^9 4l..4l'4f$ kind where $4l..4l'$ are spectator electrons, one such electron for the Cu-like sequence,two for the Zn-like,etc.... These transitions appear on the long wavelength side of the Ni-like transitions. The present work deals with some particular properties of this second kind of transitions and the interpretation of the corresponding experimental spectra. It must be stressed that although

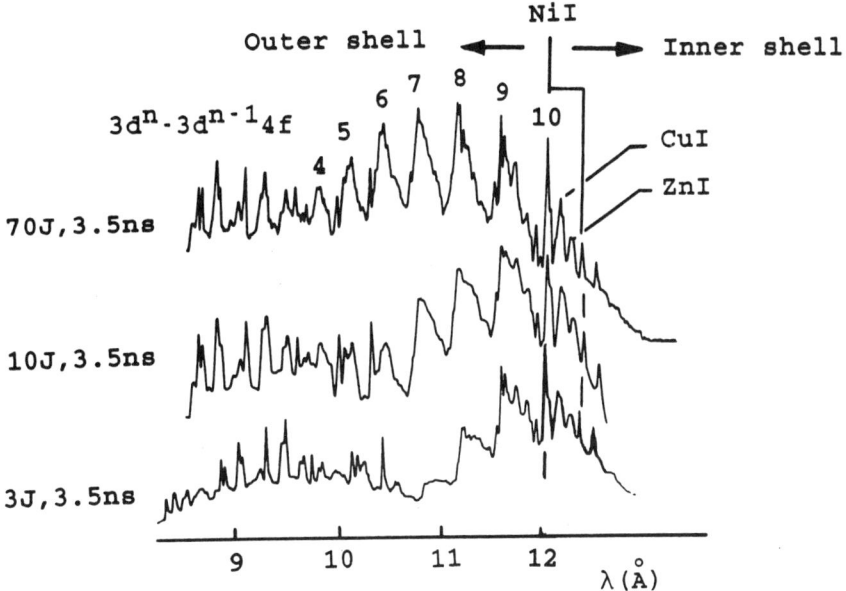

Figure 1: Spectrum of laser-produced Lanthanum plasma for different values of the laser pulse energy. (Courtasy of A. Zigler)

the present work will deal essentially with spectra emitted from laser-produced plasma, this kind of transitions has been observed in the spectra emitted by many other sources: vacuum sparks[2], exploded-wire[3], tokamak[4] and EBIT[5] sources.

SPECTROSCOPIC ANALYSIS OF THE CONTINUUM

First attempts to make a spectroscopic analysis of the 3d-4f inner-shell excited spectrum were performed by Klapisch et al.[6] for the spectra of thulium to Rhenium and by Busquet et al.[7] for the spectra of highly ionized gold. It was observed that conventional methods of line identification by parametric fitting could not be used for the identification of lines in these spectra, since most of the features were blended. Identification could rely on ab-initio calculation using relativistic codes only. The RELAC code[8] was used by Busquet et al. to give an extensive list of wavelengths, each feature in the actual spectra corresponding to many computed lines. Indeed, every transition has many close lines, which gives rise to an Unresolved Transition Array (UTA) feature. Moreover, for each ionization state, since the difference in binding energies between the 41 electrons is small compared to the temperature, one has to consider many possible transitions-e.g., for the Cu-like case $3d^{10}4l-3d^94l4f$ (l=s,p,d,f). All these transitions are nearly superimposed and not distinguishable one from the other. For these reasons Klapisch et al.[6] used the then newly developped SOSA model[9], a version of the UTA[10] model applicable when the array is split according to one or several spin-orbit interactions. The idea was to use average quantities (line-strength weighted mean wavelength λ and width $\Delta\lambda$) obtained from compact formula instead of performing laborious, time consuming full ab-initio intermediate coupling calculations that give useless lists of hundreds (or thousands) of unresolved lines. A first complication in the use of the SOSA model was found[7] to be the breakdown of the j-j coupling scheme in relevant level structures.

BREAKDOWN OF j-j COUPLING

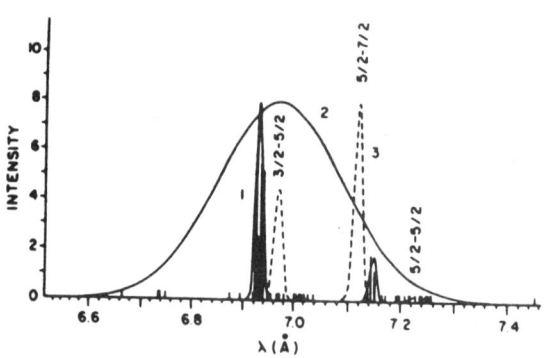

Figure 2. Comparison of different kinds of calculations for the array $3d^{10}4p$ - $3d^94p4f$ in Cu-like Tm XLI.

Figure 2 shows the results of three kinds of computations of the $3d^{10}4p-3d^94p4f$ transition in Cu-like Tm^{+40} The first curve represents the envelope of a detailed ab-initio intermediate coupling computation where each line has been given an experimental width. It shows clearly why the UTA description of the array as a single gaussian is useless (curve 2). Indeed the array is conspicuously split into subarrays corresponding to the different 3d-4f allowed jj transitions. Curve 3 shows the results of the SOSA formulas where each j-j subarray is now a separate gaussian. In this third curve, the relative intensity of the sub-arrays has been given according to pure j-j coupling. Figure 2 shows some remarkable facts: the width of of the sub-arrays as computed by the formula of the SOSA model essentially reproduce the IC single configuration computation, although the wavelengths are shifted and the relative intensities are quite different from the pure j-j coupling relative intensity. The origin of the discrepancy between curves 2 and 3 is the breakdown of the pure j-j coupling scheme in the actual IC computations. However, it was shown that the correction for the wavelength shift of the subarray was not in a first order approximation a function of the number of spectator electrons and, using an adequate correction for this wavelength shift,the theoretical results could be compared to the available experimental spectra. At the present time, the correction of the SOSA intensities in the case of breakdown of j-j coupling has been introduced in the SOSA model[11]. A comparison between an experimental spectrum of highly ionized Thulium (Z=69) is given in Fig. 3. This analysis shows that about four ionization states (from the Cu-like to the Ge-like sequence) were responsible for the 3d-4f inner shell emission in the spectra of the elements Tm, Yb, Hf and Ta. This was confirmed by the independant analysis of the 3d-5f transitions in a quite similar Ta laser -produced plasma[12]. In this latter case the analysis seems to be easier since the 3d-5f arrays are much closer to pure j-j coupling. Also ionization states are very well separated and the experimental spectrum does not degenerate into a pseudocontinuum.

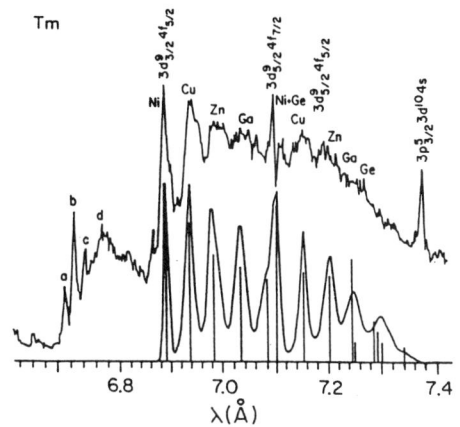

Figure 3: Experimental and theoretical spectrum of highly ionized Tm (ref.6)

The conclusion is that the SOSA model is a very efficient tool to determine the number of ionisation states emitting the 3d-4f pseudocontinuum. Some discrepancies remain however: the predicted SOSA width seems too small (See fig. 3).

CONFIGURATION MIXING.

The problem of the discrepancy between theoretical and observed 3d-4f pseudocontinuum calculations was addressed by Busquet et al.[7,13] and Bauche-Arnoult et al.[14]. Busquet et al. (ref. 7) showed that to reproduce the experimental spectrum of Au , a "red wing" had to be added to the line profile. Better agreement was also obtained by accounting for cascades and $\Delta n=0$ transitions[13]. Bauche-Arnoult et al. suggested that significant population of the $3d^94f4l4l'..5l*$ levels could be built through capture of free electrons into these autoionizing configurations. Radiative decay to the $3d^{10}4l4l'..5l*$ lower levels could result in the emission of SOSA slightly shifted towards shorter wavelength. These arrays could fill the empty

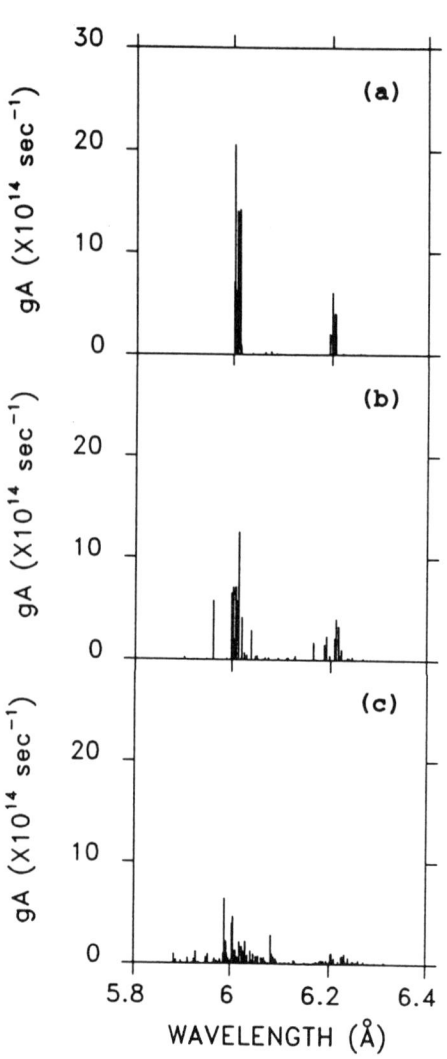

Figure 4. Transitions in Ga-like Ta XLIII:
(a) $3d^{10}4s^24p-3d^94s^24p4f$.
(b) $3d^{10}4s^24p-3d^9[4s^24p4f+4s^24d^2]$
(c) $3d^{10}4s^24p-3d^9[4s^24p4f+4s^24d^2+4s4p^24d]$

space between the 3d-4f transitions with 41 spectator electrons only. Intermediate coupling calculations show, however that configuration mixing can account for this discrepancy even if only resonance arrays excited by electron impact from the ground state are taken into account. The global effect of the configuration mixing on 3d -4f transitions will be shown here only for the case of Ga-like $3d^{10}4s^2-3d^94s^24p4f$ transition in tantallum although similar calculations have been performed for Cu- and Zn-like tantallum. In the theoretical spectra of figure 4, the height of each line is proportional to g x A , g being the statistical weight of the upper level and A the transition probability gA . The first theoretical spectrum shows the results of the computation before introduction of the mixing. The second shows the results of the computation including the perturbing configuration. The following conclusions can be drawn from the Figures 4 and are in fact general for the case of transition with the possibility of 4s4f-4p4d or 4p4f 4d² interaction:
i) The general structure of three subarrays is not changed.
ii) The subarray widths are much larger when configuration mixing is introduced
iii) Line intensities are smaller in the case of configuration mixing because this intensity is shared between many more lines.
iiii) It seems that the mean wavelengths of the subarrays are not changed.

To express these facts in a more quantitative way, the line strength-weighted mean wavelengths λ and spectral widths $\Delta\lambda$ of the arrays were calculated using the computed data for individual lines (unlike the UTA or SOSA models) . λ and $\Delta\lambda$ are defined by the following equations:

$$\bar{\lambda} = \frac{\sum_{i,j} g_i A_{ij} \lambda_{ij}}{\sum_{i,j} g_i A_{ij}} \quad , \quad \Delta\lambda = \left\{ \frac{\sum_{i,j} (\lambda - \lambda_{ij})^2 g_i A_{ij}}{\sum_{i,j} g_i A_{ij}} \right\}^{1/2}$$

where λ_{ij} is the wavelength , A_{ij} the Einstein coefficient of the computed transition line ,and g_i is the statistical weigth of the upper level . Only those lines in the selected wavelength range r are included in the summation. λ represents the line-strength weighted mean wavelength and $\Delta\lambda$ the spectral width of the subarrays (these results are equivalent to the results of the SOSA model for the single configuration calculations). The weighted mean wavelengths and spectral widths obtained for the $3d^{10}4s-3d^9[4s4f+4p4d]$ transition are given in Table 1. The mean wavelength and spectral width has been computed for lines in three different wavelength ranges r_1, r_2 and r_3 corresponding to the maximum spread of the subarrays. From Table 1,the very significant effect of configuration mixing on the SOSA width is evident. This shows that results of computations of the width of the 3d-4f arrays using the SOSA model should be taken with caution. On the other hand,it appears from Table 1 that the mean wavelength of the two intense subarrays is almost not changed.

Wavelength range	[0,6.107]		[6.107,6.246]		[6.246,∞]	
Transition	λ(Å)	$\Delta\lambda$(Å)	λ(Å)	$\Delta\lambda$(Å)	λ(Å)	$\Delta\lambda$(Å)
$3d^{10}4s^24p-3d^94s^24p4f$	6.0078	9.2	6.2029	5.2	6.2715	39.3
$3d^{10}4s^24p-3d^94s^2[4p4f+4d^2]$	6.0066	22.2	6.1996	25.0	6.2790	29.7
$3d^{10}4s^24p-3d^9[4s^24p4f+4s^24d^2+4s4p^24d]$	6.0085	44.9	6.1998	33.9	6.2897	58.8

Table 1. Mean Wavelength and width of tantallum Ga-like 3d-4f transition subarrays calculated from the results of full intermediate coupling line computation for lines in three different wavelength ranges r_1, r_2 and r_3

IONIZATION STATES CONTRIBUTING TO 3d-4f EMISSION

Since the mean wavelength of the 3d-4f SOSA subarrays is not affected by the configuration mixing, this remains an effective tool to determine the number of ionization states contributing to the 3d-4f pseudocontinuum. Following this conclusion, a review of the M-shell published spectra emitted from laser-produced plasma of elements with Z=42 (Mo)to Z=92 (U) was undertaken. This review did not include spectra emitted by very high density compressed or femtosecond pulse laser-produced plasmas and a common property of the emitting plasma is the rather low 10^{20}-10^{21}cm^{-3} electron density. This review led to the following conclusions :

1. The number of SOSA appearing in the spectrum of each element shows a very definite Z dependance : for relatively low Z elements (i.e Molybdenum) only two SOSA pertaining to the Cu and Zn-like sequence have, to our best knowledge, ever been observed. When analysing M-shell spectra of higher-Z element, the number of these SOSA gradually increases and in the uranium spectra as many as six SOSA can be identified in some M-shell spectra.

2. This number of SOSA seems to be quite independant of the conditions of iradiation of the target for low to intermediate Z.

Figure 1 shows that when lowering the energy of the impinging laser pulse on the La target, there is a very definite change in the emission of the outer-shell 3d-4f transitions. The inner shell pattern, however, seems to be quite independant of the power of the laser pulse. This can be observed also in other similar experiments. On the basis of these conclusions the structure of the M-shell 3d-4f spectra has been correlated to the Z dependance of the ratio of the average energy of the 3d-4f ground excited transition to the ionization energy. Figure 5 shows a plot of the ratio of the average energy of the 3d-4f transition excited from the ground for five isoelectronic sequences (Cu- to As-like). It can be seen that this ratio is a decreasing function for Z for every sequences. For low Z values this ratio is greater than 1, meaning that the excited configuration is autoionizing. For higher Z, this ratio decreases progressively and for each isoelectronic sequences crosses the value 1 at a different value of Z.

COMPARISON WITH EXPERIMENT.

From our survey of the published spectra of many high Z elements, it appears that there is a clear correlation between the number of ionization states identified in the 3d-4f spectrum and the relative position of these excited levels and the ionization potential. For example, in the spectrum of lanthanum (Z=57, Figure 1), only the Cu- and Zn-like ion have 3d-4f transitions lower than the ionization limit (Figure 5). This corresponds to the SOSA observed in the experimental

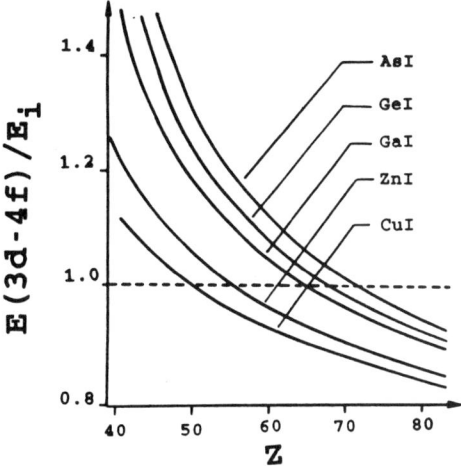

Figure 5: Ratio of the average energy of the $3d^{10}4s^{1}4p^{m}$ - $3d^{9}4s^{1}4p^{m}4f$ transition to the first ionisation potential for the Cu- (l=1,m=0), Zn- (l=2,m=0), Ga-, Ge- and As-like (l=2,m=1 to 3) isoelectronic sequences

spectrum. For higher Z elements such as Thulium (Z=69 ,Figure 3) four ionization states are identified in the experimental spectrum corresponding to the four ionization states whose 3d-4f levels are lower than the ionization limit. Clearly, a detailed multiple ion collisionnal-radiative model is needed to assess the importance of the opening of ionization channels on the intensity of the 3d-4f transitions. However, results in the simpler coronal limit will be presented here. The opening of the ionization can change, in this approximation, the intensity of 3d-4f array through two processes:
First, this can enhance the ionization rate and shift the ionization balance towards higher ionization states, because of the opening of excitation-autoionization (EA) channel. This has been confirmed by calculations of the EA rates using the HULLAC code for 3d-4f transition along the Ga-like isoelectronic sequence. These calculations have been performed by the HULLAC team at Beer-Sheva and Lawrence Livermore. These calculations show, for example, that in the case of highly ionized Praseodymium, the introduction of the EA process through 3d-4f transition can lower the electron temperature of most abundance Ga-like ion by as much as 1/3 . It will only be

stressed that in the case of EA processes, 3d-4d transitions are also important, but their isoelectronic behaviour is very close to the one of the 3d-4f transition.

Second, the opening of autoionization channels can lower the intensity of the 3d-4f transitions because of the autoionization branching ratio. When the direct inner-shell excitation is the principal excitation mechanism, the line intensity is proportionnal to the fluorescence yield defined by:

$$Y_{if} = \frac{A_{if}^R}{\sum_l A_{il}^A + \sum_k A_{ik}^R}$$

The lowering of the intensity of the particular $f \rightleftarrows i$ transition because of the opening of the autoionization channel is :

$$\rho_{if} = \frac{\sum_k A_{ik}^R}{\sum_l A_{il}^A + \sum_k A_{ik}^R}$$

One can define an intensity-weighted average of ρ_{ij} on all the transitions :

$$\rho = \frac{\sum_i \sum_j \rho_{ij} A_{ij}^R}{\sum_i \sum_j A_{ij}^R}$$

Table 2 gives the results of calculations of ρ for the $3d^{10}4s^24p - 3d^9[4s^24p4f + 4s^24d^2]$ transitions for some of the ions of the Ga-like sequence.

ion	Mo XII	Ag XVII	La XXVIII	Ta XLIII
ρ	0.017	0.08	0.31	0.80

Table 2: Computed value of ρ for selected ions of the Ga-like sequence.

For the lower Z member of the isoelectronic sequence the intensity of the array can be lowered by a factor as high as 100.
This can explain the isoelectronic trend of the 3d-4f observed spectra.

ACKNOWLEDGEMENT

The author of is gratefull to the Solar Terrestrial Relationships Branch of the E.O. Hulburt Center for Space Research at the Naval Research Laboratory for hospitality during his stay as a guest scientist. This work was performed in collaboration with A. Bar-Shalom and J. Oreg of the Negev Nuclear Center,J.F. Seely at the Naval Research Laboratory, J.L. Schwob and D. Mitnik of the Hebrew University ,M. Klapisch of Lawrence Berkeley Laboratory and W. Goldstein at Lawrence Livermore Laboratory.

*Permanent address: Racah Institute of Physics, The Hebrew University, Jerusalem, Israel

REFERENCES

[1] P.G. Burkhalter,D.J. Nagel and R.R. Whitlock, Phys. Rev. **A9**,2331(1974).

[2] J.L. Schwob,M. Klapisch,N. Schweitzer,M. Finkenthal,C. Breton, C. De Michelis and M. Matioli,Phys. Lett.**A62**,85(1977).

[3] P.G. Burkhalter,C.M. Dozier and D.J. Nagel, Phys. Rev. **A15**, 700(1977)

[4] S. von Goeler, P. Beiersdorfer, M. Bitter, R. Bell,K. Hill, P. LaSalle,L. Ratzan , J. Stevens, J. Timberlake, S. Maxon and J. Scofield , J. de Physique,Colloque C1,Supplement au n°3, **49**,181(1988).

[5] N.K. Del Grande,P Beiersdorfer,J.R. Henderson,A.L.Osterheld, J.H Scofield and J.K. Swenson, Eleventh International Conference on the Application in Research and Industry, Denton,Texas, september 1990.

[6] M. Klapisch,P. Mandelbaum , A. Zigler,C. Bauche-Arnoult, and J. Bauche ,Phys. Scripta **34**,51(1986).

[7] M. Busquet,D. Pain,J.Bauche and E. Luc-Koenig,Phys. Scr. **31**, 137(1985).

[8] M.Klapisch,J.L. Schwob,B.S. Fraenkel and J. Oreg,J. Opt. Soc. Am. **67**,148(1977).

[9] C. Bauche-Arnoult,J. Bauche and M. Klapisch,Phys. Rev **A31**, 2248(1985).

[10] C. Bauche-Arnoult,J. Bauche and M. Klapisch,Phys. Rev **A20**, 2424(1979) and J. Bauche,C. Bauche-Arnoult and M. Klapisch, Adv. At. Mol. Phys.,**27**,131(1987).

[11] J. Bauche,C. Bauche-Arnoult and M. Klapisch,J. Phys.**B24**, 1(1991).

[12] P. Audebert,J.C. Gauthier,J.P. Geindre,C. Chenais-Popovics, C.Bauche-Arnoult,J. Bauche,M. Klapisch,E. Luc-Koenig and J.F. Wyart,Phys. Rev **A32**,409(1985).

[13] M. Busquet and J. Bruneau, J. de Physique,Colloque C6, Supplement au n°10,47,333(1986).

[14] C. Bauche-Arnoult,J. Bauche,E. Luc-Koenig,R.M. More,J.F. Wyart C. Chenais-Popovics,J.C. Gauthier,J.P. Geindre and N. Tragin, Phys. Rev. A39,1053(1989).

BOUND STATES AND IONIZATION KINETICS IN DENSE PLASMAS

W. Ebeling, I. Leike, U. Leonhardt
Institute of Theoretical Physics, Humboldt-University
Invalidenstr. 42, D-1040/O Berlin, Germany

ABSTRACT

Bound states in dense plasmas show strongly density-dependent characteristics. First the shifts of the energy levels and the corresponding influence on the ionization equilibria are analyzed. The stability analysis of the thermodynamic functions shows the existence of additional phase transition at high temperatures and high pressures due to nonideality effects. The nonideality also influences the kinetic rate coefficients and changes the time behaviour of the electron populations of the bound states.

INTRODUCTION

Investigations of ionization processes in dense plasmas are of interest for the study of stars and planets as well as for technical applications connected e.g. with near electrode plasmas in arcs or with plasmas produced by high-power laser beams. Most of the existing theories are restricted so far to hydrogen and hydrogen-like plasmas [1] where quantum-statistical methods are well elaborated, and, on the other hand to plasmas of the heavy elements, where the Thomas-Fermi theory yields an excellent approximation. Due to sincere conceptual and computational difficulties the number of investigations devoted to plasmas of elements of the 2nd and the 3rd period are still rather limited. The paper gives a survey of the state of the theory with respect to hydrogen and the light elements. In the first part of this paper the known analytical results for the bound states, the thermodynamic properties and the degree of ionization are given. Starting from the Bethe-Salpeter equation first the density dependence of the energy levels is discussed.

Then the thermodynamic functions and the ionization equilibrium are analyzed. We start with the consideration of pure hydrogen plasmas and pay special attention to the existence of phase transitions. It is shown that in systems with long-range coulombic interactions besides the classical first-order transition typical for neutral gases a second first-order transition may appear [2]. Along the coexistence line a system undergoing a plasma transition is divided into two phases of different density and different degree of ionization. The experimental and theoretical efforts to check the existence of this transition are discussed.

The third part of the paper is devoted to the discussion of kinetic transitions. The time evolution of the relative populations of excited atomic levels in reacting H-like dense plasmas are studied on the basis of rate equations. The many particle effects are taken into account via energy level shifts. Numerical results are presented for hydrogen, helium and carbon.

IONIZATION EQUILIBRIUM AND THERMODYNAMIC PROPERTIES

THE ENERGY LEVEL SHIFTS

Bethe and Salpeter developed a generalization of the Schrödinger equation which yields the energy levels and the wave functions of bound states embedded in a medium. In momentum representation the Bethe-Salpeter-equation reads

$$[E(p_1) + E(p_2) - z]\,\psi(p_1,p_2,z) + \sum_q V(q)\,\psi(p_1+q, p_2-q, z)$$

$$+ \sum_q V^{pl}(p_1,p_2,q,z)\,\psi(p_1+q, p_2-q, z) = 0. \tag{1}$$

In a hermitian approximation the effective energy levels may be represented by

$$\bar{E} = E_k + \Delta_e + \Delta_+ + \Delta_k. \tag{2}$$

Here Δ_e and Δ_+ are the selfenergies of the electrons and their positively charged partners and Δ_k is the so-called level shift which decreases with increasing quantum number k. In the continuum limit we have

$$\bar{E}_\infty = \Delta_e + \Delta_+. \tag{3}$$

This shows us, that the effective binding energies, i.e. the differences between the level and the continuum (the binding energies) are given by

$$\bar{E}_k = E_k + \Delta_k. \tag{4}$$

In dense plasmas the number of levels is always finite. The principal schema is shown in Fig.1.

In our calculations we use Padé or polynomial approximations for the level shifts, which are based on static calculations. An example is the following formula which is a simplified variant of the polynomial given by EBELING and LEIKE [3]

$$\Delta_k = ze^2\kappa(1 - a_k\kappa + b_k\kappa^2 - c_k\kappa^3 - f_k\chi^4); \tag{5}$$

$$f_k = \chi^{-4}(1 - a_k\chi_k + b_k\chi_k^2 - c_k\chi_k^3 - \frac{1}{2}k^{-2}\chi_k^{-1}), \qquad \chi = \kappa a_B. \tag{6}$$

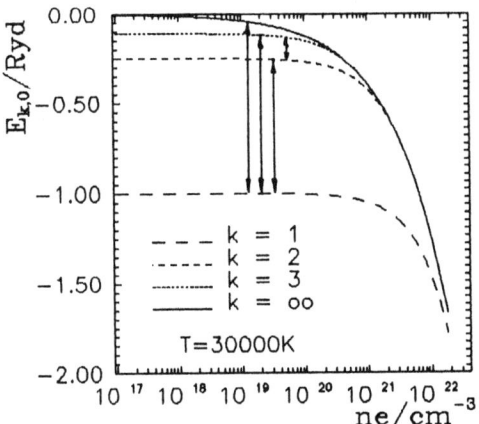

Figure 1: Schema of the hydrogenic levels and the transitions

The χ_k denote the values of (κa_B) where the level k merges into the continuum. These values are known from numerical solutions of the Schrödinger equation [4]. In the static approximation one finds

$$a_k = \frac{1}{2} <\rho>_k, \qquad \rho = r/a_B, \tag{7}$$

$$b_k = \frac{1}{6} <\rho^2>_k, \qquad c_k = \frac{1}{24} <\rho^3>_k. \tag{8}$$

The approximation (5-8) is sufficiently accurate in most cases if one is interested in the shifts themselves. It reflects the Debye shifts as well as the disappearance of the levels at higher densities. However if we are looking for differences between two bound states, dynamic calculations have to be used [5,6]. The basic result of such dynamic calculations are formulae linear in the density

$$\Delta_k - \Delta_m = A_{km} n_e, \tag{9}$$

where the A_{km} are certain numerical coefficients give by the above mentioned authors. However the numerical coefficients A_{km} are not consistent with eqs.(5-8). In other words, a dynamical recalculation of the coefficients a_k which is consistent with (9) is required, as well as a recalculation of the χ_k, b_k and c_k. In spite of this we shall use eqs.(5-6) having in mind the necessity of a more accurate recalculation of the coefficients.

THERMODYNAMICAL EFFECTS OF NONIDEALITY

Defining the population densities of electrons in free states by $n_e = n_\infty$ and in bound states by n_k we get a Saha equation for the determination of these quantities

$$\frac{n_k^0}{n_e^0 n_+^0} = k^2 \Lambda^3 \exp(E_k/k_B T). \tag{10}$$

Since the effective binding energies disappear with increasing density due to nonideality, we observe a density ionization [2].

At low densities where the shifts of the levels are still negligible we observe the standard behaviour known from solutions of the ideal Saha equation, i.e. one observes first full ionization of both electrons $z = 2$, then with increasing density ionic bound states $z = 2$. Finally in the region $n \approx 10^{19} - 10^{22} cm^{-3}$ atomic bound states $z = 2$ are observed. Any further increase of the density leads to a sudden decrease of the binding energies corresponding to eqs.(5-6). Due to this nonideality effect, first the atomic bound states and then the ionic bound states break down leading finally to full ionization again. At very large density bound states may not persist.

Another type of solutions of the Bethe-Salpeter equation leads to the two-particle Greens functions which may be used for the determination of thermodynamic functions including nonideality effects [2]. For practical calculations we have developed Padé approximations for the thermodynamic potentials. Instead of going into details we refer only to some references [7,8]. Let us discuss here only the most dramatic nonideality effect, namely the appearance of additional phase transitions due to Coulomb interactions, which was predicted first by LANDAU, ZELDOVICH, NORMAN and STAROSTIN and calculated recently in several theoretical papers [1,9,10]. This transition is so to say a discrete form of the density ionization discussed above.

We start with the consideration of pure hydrogen plasmas. In the above mentioned work the thermodynamical stability criteria for systems with long-range coulombic interactions were checked. It was shown that besides the classical first-order transition typical for neutral gases a second first-order transition may appear [2]. Along the coexistence line a system undergoing a plasma transition is divided into two phases of different density and different degree of ionization. There seems to be experimental and theoretical evidence that in metals the van der Waals transition and the plasma phase transition fuse and that there remains a transition which is a mixture of both types. However in hydrogen and in hydrogen-like and in helium-like systems the van der Waals transition and the plasma phase transition seem to be well separated. Our theory yields the estimate for the second critical point of hydrogen

$$C_2 : T = 16500\,K, \quad p_c = 22.8\,GPa, \quad \rho_c = 0.13\,gcm^{-3}$$

The coexistence line and the lines of constant ionization have been determined. It was shown that around the critical point C_2 a very quick change in the composition occurs. We also note that near the critical point the coexisting gases and liquids are only partially ionized (about 30 may coexist with a metallic liquid. One of the most important results obtained so far is the drastic lowering of the pressures for the transition to metallic hydrogen in comparison with the corresponding transition pressure in the solid state, which is typically in the range $200-800\,GPa$. The expected pressures for the transition from liquid molecular hydrogen to liquid metallic hydrogen are ten times smaller than those for the transition from solid molecular hydrogen to solid metallic hydrogen. This may be of interest for laboratory comparisons with adiabatic compression of hydrogen as well as for astrophysical applications. Following our estimates the metallization of the interior of Jupiter occurs at much lower pressures, i.e. nearer to the surface of the planet, than assumed in earlier estimates. In order to find the metal non-metal transition in hydrogen at laboratory conditions very complicated experiments are necessary. Recent static experiments with hydrogen at $77\,K$ give some evidence that the metallization occurs at $200\,GPa$. In order to study the transition at higher temperatures one should start adiabatic compression experiments with initial conditions near to the jovian adiabate. GRIGORIEV et al. [12] report about a density step from $1.1\,gcm^{-3}$ to $1.3\,gcm^{-3}$ at the pressure of about $280\,GPa$ and temperatures of about $10^4\,K$. Further NELLIS et al. [13] and ROSS et al. [14] report about two pressure data at the highest densities (about $0.5\,gcm^{-3}$) which are decreasing with the density. The experimental findings could well be a first hint to the existence of a plasma phase transition in hydrogen. So far however there are no reliable data which could be considered as an experimental evidence. Further theoretical and experimental work is necessary to decide the open question of the existence of a plasma phase transition in hydrogen.

The calculations for Helium plasmas show the splitting of C_2 into two critical points, connected with single and double ionization respectively as shown in recent work by FÖRSTER et al. [11]. This corresponds to the two step-like transitions we observe at the highest densities in Fig. 2. Our estimate gives the critical data,

$$C_2' : T_c \cong 35000\,K, \quad p_c \cong 0.7\,TPa$$
$$C_2'' : T_c \cong 120000\,K, \quad p_c \cong 10\,TPa$$

However these are only qualitative estimates since the calculation of critical points requires very high accuracy of the thermodynamic functions.

KINETICS OF LEVEL POPULATIONS

MASTER EQUATION AND TRANSITION PROBABILITIES

In equilibrium we have obtained the electron distribution on of energy spectrum by solving a set of Saha equations. In nonequilibrium we assume that the distribution is given by a set of kinetic equations

$$\frac{dn_k}{dt} = \sum_{k'} W_{kk'} n_{k'} - \sum_{k'} W_{k'k} n_k. \qquad (11)$$

Here the rates are given by

$$W_{k'k} = C_{k'k} + R_{k'k}. \qquad (12)$$

The first contribution describes the transitions caused by electron collisions and the second contribution the transitions due to interactions with photons. We consider here only the collisional transitions. Since exact quantum-statistical results including nonideality are available only in a few cases [15], we have to introduce several approximations.
- the Born approximation for the scattering processes is used
- no screening is taken into account in the matrix elements
- shifted binding energies corresponding to (4) are used
- the wave functions are assumed to be Coulombic

With these approximations the collision rate coefficients are given by

$$C_{\infty k} = \int \frac{d^3p}{(2\pi\hbar)^3} \frac{d^3p'}{(2\pi\hbar)^3} \frac{d^3\bar{p}}{(2\pi\hbar)^3} \frac{2\pi}{\hbar} \delta(\frac{p^2}{2m} + \bar{E}_k - \frac{p'^2}{2m} - \bar{E}_p)$$

$$\times \sum_{lm} |<p'\bar{p}|V|klmp>|^2 \frac{1}{k^2}\Lambda^3 \exp\left(-\frac{p^2}{2mk_BT}\right), \qquad (13)$$

$$C_{k'\infty} = \int \frac{d^3p}{(2\pi\hbar)^3} \frac{d^3p'}{(2\pi\hbar)^3} \frac{d^3\bar{p}}{(2\pi\hbar)^3} \frac{2\pi}{\hbar} \delta(\frac{p^2}{2m} + E_k - \frac{p'^2}{2m} - E_p)$$

$$\times \sum_{l'm'} |<k'l'm'p'|V|\bar{p}p>|^2 n_+ \Lambda^6 \exp\left(-\frac{p'^2}{2mk_BT} - \frac{E_p}{k_BT}\right), \qquad (14)$$

$$C_{k'k} = \int \frac{d^3p}{(2\pi\hbar)^3} \frac{d^3p'}{(2\pi\hbar)^3} \frac{d^3\bar{p}}{(2\pi\hbar)^3} \frac{2\pi}{\hbar} \delta(\frac{p^2}{2m} + \bar{E}_k - \frac{p'^2}{2m} - \bar{E}_p)$$

$$\times \sum_{lml'm'} |<k'l'm'p'|V|klmp>|^2 \frac{1}{k^2}\Lambda^3 \exp\left(-\frac{p^2}{2mk_BT}\right). \qquad (15)$$

In correspondence to the approximations formulated above, V is here the bare potential. In equilibrium detailed balance for each transition should be observed what leads to the set of Saha-type equations (10) and the Boltzmann distribution

$$\frac{n^0_{k'}}{n^0_k} = \frac{k'^2}{k^2} \exp\left(-\frac{E_{k'} - E_k}{k_B T}\right). \quad (16)$$

The advantage of the approach described above is, that all results known for the pure Coulombic case remain valid with little modifications which basically are connected with the replacement of the Coulombic energies E_k by the new binding energies E_k. The procedure of calculations will be demonstrated now for a simple case.

ONE-STATE HYDROGEN SYSTEMS

The simplest kinetic model is a hydrogen plasma with only one bound state - the ground state - and no radiative transitions. Then the master equations reduce to just one ordinary differential equation

$$\frac{dn_e}{dt} = \alpha(n - n_e)n_e - \beta n_e^2; \qquad \alpha = C_{\infty 1}, \qquad \beta = C_{1\infty}. \quad (17)$$

The equilibrium condition is

$$\alpha = \Lambda^3 \exp(I/k_B T), \qquad I = |E_1 + \Delta_1|. \quad (18)$$

In the approximation introduced above the ionization coefficient is

$$\alpha = \frac{8\pi m}{\hbar^3} \Lambda^3 \int_I^\infty d\left(\frac{p^2}{2m}\right) \sigma(p) \frac{p^2}{2m} \exp\left(-\frac{p^2}{2mk_B T}\right), \quad (19)$$

$$\sigma(p) = \frac{m}{p} \int \frac{d^3 p'}{(2\pi\hbar)^3} \frac{d^3 \bar{p}}{(2\pi\hbar)^3} \frac{2\pi}{\hbar} \delta\left(\frac{p^2}{2m} - I - \frac{p'^2}{2m} - \frac{\bar{p}^2}{2m}\right) |<p'\bar{p}|V|p1>|^2. \quad (20)$$

A standard approximation for the scattering section of ideal plasmas is the BIBERMANN formula [16]

$$\sigma(p) = 2.5\pi a_B^2 \frac{p^2}{2m|E_1|} \ln\left(\frac{p^2}{2m|E_1|}\right). \quad (21)$$

In order to introduce nonideality corrections SCHLANGES BORNATH and KREMP [15] took into account that the δ-function in eq. (19) contains the shifted argument

$$p^2/2m + E_l + \Delta_l. \quad (22)$$

This lead the above mentioned authors to the the approximation

$$\alpha(p) = \frac{a_B^2 Ry}{p^2/2m} 2.5\pi \ln\frac{p^2/2m + \Delta}{|E_1|} \quad (23)$$

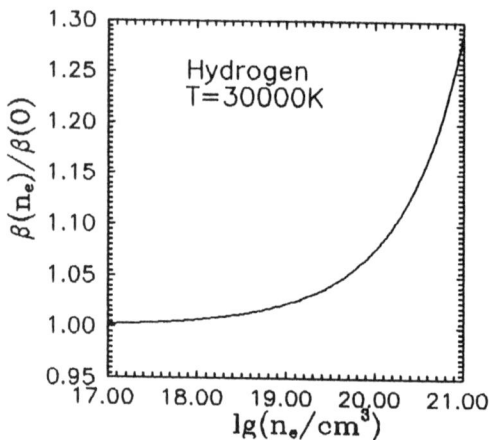

Figure 2: Density dependence of the recombination coefficient

and finally to

$$\alpha = \alpha_{id} \exp(\Delta/kT), \qquad \beta = \beta_{id}. \qquad (24)$$

This corresponds to an Arrhenius behaviour of the the rates, i.e. upward transitions increase exponentially with energy shifts and downward transitions remain unchanged. This assumption is physically plausible and therefore it was used also in many of our concrete calculations [3]. However we did also some efforts to improve the approximation (24). Our numerical calculations of the original expression (19) show that an actual improvement of eq. (23) is given by

$$\alpha(p) = \frac{a^3 Ry}{p^2/2m}\, 2.5\pi\, \ln(\frac{p^2}{2m\,I}), \qquad I = |E_l + \Delta|. \qquad (25)$$

In other words, the energies and the shifts remain unseparated. This approximations yields the expressions

$$\alpha = 10 a_B^3\, \Lambda\, \frac{Ry}{2\pi\hbar}\, \mathrm{Ei}(-I/k_B T),$$

$$\beta = 10 a_B^3\, \Lambda^4\, \frac{Ry}{2\pi\hbar}\, \mathrm{Ei}(-I/k_B T)\, \exp(I/k_B T). \qquad (26)$$

These expressions are identical with the results for Coulombic interactions, except that the bare ground state energy I_0 is replaced by the new effective binding energy I which includes the shift. Asymptotically, i.e. for $I \gg kT$ the new expressions are identical with eqn.(24). At higher temperatures $I \sim kT$ however, differences are observed. Especially β is getting density-dependent and increases with the density (Fig.2).

MANY-STATE HYDROGEN-LIKE SYSTEMS

So far no quantitative theory is available which describes nonideal many-state systems. Generalizing the Eyring-type behaviour of the ground state coefficients (24) we have proposed in earlier work the following set of coefficients for all (collision-induced) rates

$$C_{\infty k} = \alpha_k = \alpha_k^{id} \exp(\Delta_k/kT),$$
$$C_{k\infty} = \beta_k = \beta_k^{id},$$
$$k > m: \quad C_{km} = \alpha_{km} = \alpha_{km}^{id} \exp[(\Delta_k - \Delta_m)/kT],$$
$$k < m: \quad C_{km} = \beta_{km} = \beta_{km}^{id}. \quad (27)$$

The principle is, that uphill transitions get the Eyring factor whereas downhill transitions do not. This assumption yields a good approximation at least for far transitions $| E_k + \Delta_k - E_m - \Delta_m | \gg k_B T$. For near transitions we are proposing here to modify eqn.(27) by using the recipe that we have replaced the bare binding energies in the Coulombic expressions by the shifted energies $\bar{E}_k$. This recipe we have demonstrated for the one state model by introducing the shifts into the argument of the Ei-functions. Let us still note that there exist several interesting scaling properties with respect to the z-dependence which allow us to extend the theory from hydrogen $z = 1$ to higher charges $z > 1$ as helium and carbon, what will be used also below. In brief the scaling properties are:

$$C_{k'k}^z = z^{-3} C_{k'k}^l (k_B T/z^2, \kappa/z), \qquad R_{k'k}^z = z^4 R_{k'k}^l (kT/z^2, \kappa/z). \quad (28)$$

THE TIME BEHAVIOUR OF ELECTRON POPULATIONS

We have given above a complete set of transition rates. In the following we give some applications to hydrogen and to hydrogenlike helium and carbon plasmas, restricting us to the collisional transitions only. In order to get the explicit time behaviour one has to solve the set of differential equations [3]

$$\frac{dn_k}{dt} = -\alpha_k n_k n_e + \beta_k n_i n_e^2$$
$$- \sum_{m=k+1}^{nmax} \alpha_{km} n_k n_e - \sum_{m=1}^{k-1} \beta_{km} n_k n_e + \sum_{m=1}^{k-1} \alpha_{mk} n_m n_e + \sum_{mk+1}^{nmax} \beta_{mk} n_m n_e, \quad (29)$$

where $nmax$ is the maximal state number at the given density. The coefficients for the ideal case are given e.g. by DRAWIN and EMARD [17]. At time $t = 0s$ a completely ionized plasma is assumed. We studied the capture of the first electron. The capture of the second electron for ions with $z > 1$ has been

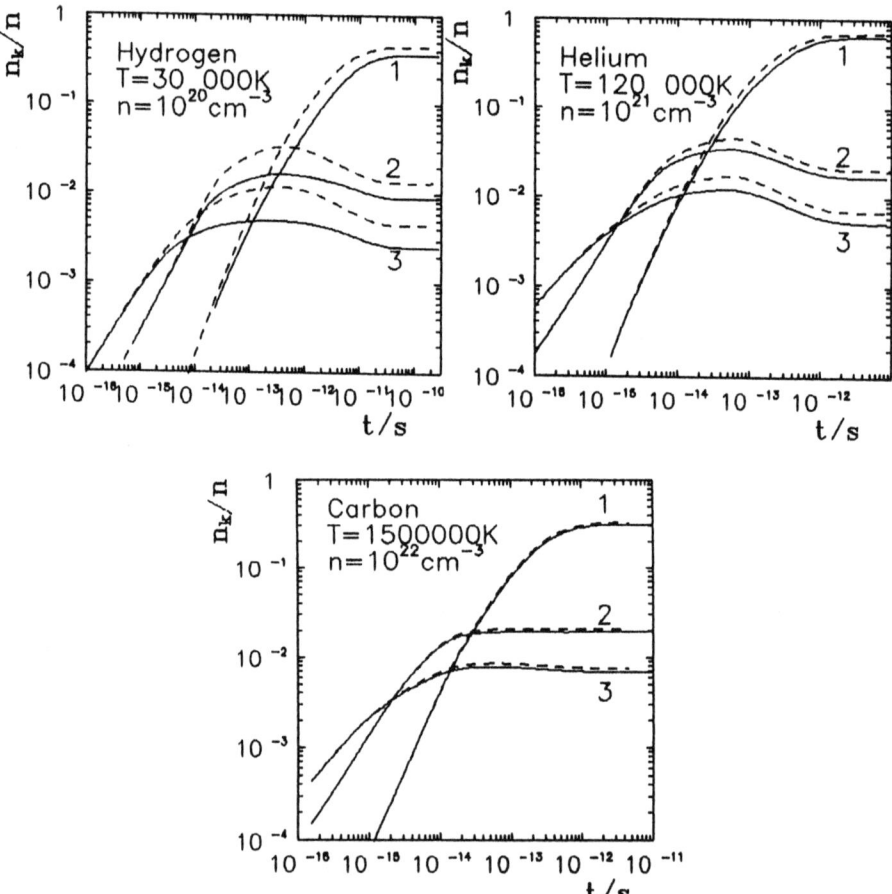

Figure 3: Time evolution of the relative population densities n_k/n of the ionic levels. Dashed curves - plasma without interaction ($\Delta = 0$), solid curves - nonideal plasma

neglected. The result is shown in Fig.3. The total density n is chosen such that a certain number of levels $nmax$ is left (in Fig.3 $nmax = 3$). The population density of nonideal hydrogen is about 3 times smaller than the ideal one. In contrast this is negligible for carbon. The nonideality is lost for carbon because of the high temperature. We have chosen the indicated temperatures to ensure, that the equilibrium density of the $(z - 1)$-fold charged ions is small (<10% of the total density). However, for lower temperatures strong nonideal effects are expected.

References

[1] W. Ebeling, W. Richert, Phys. Lett. 108A, 80 (1985).

[2] W.D. Kraeft, D. Kremp, W. Ebeling, G. Röpke, Quantum Statistics of Charged Particle Systems (Akademie-Verlag, Berlin, 1986).

[3] W. Ebeling, I. Leike, Physica 170A, 682 (1991).

[4] F.J. Rogers, H.C. Graboske, D.J. Harwood, Phys. Rev. A1, 1577 (1970).

[5] L. Hitschke, G. Röpke, Phys. Rev A37, 4991 (1988).

[6] S. Günter, L. Hitschke, G. Röpke, Phys. Rev. A15, (1991), in press.

[7] W. Ebeling, Contr. Plasma Physics 29, 165 (1989).; 30, 553 (1990); W. Ebeling, Z. phys. Chem. (Leipzig) 271, 233 (1990).

[8] K. Kilimann, W. Ebeling, Z. Naturforsch. 45a, 613 (1990).

[9] W. Ebeling, A. Förster, W. Richert, H.Hess, Physica 150A, 159 (1988).

[10] D. Saumon et al., J. Chem. Phys. 90, 7395 (1989); D. Saumon, G. Chabrier, Phys. Rev. A15, (1991) in press.

[11] A. Förster, T.Kahlbaum , W.Ebeling, XX. ICPIG Contributed Papers (Pisa 1991), Vol.2, p.385.

[12] F.V. Grigoriev et al., Zh. Eksp. Teor. Fiz. (USSR) 69, 743 (1975).

[13] W.J. Nellis et al., Phys. Rev. Lett. 48, 816 (1982).

[14] M. Ross et al., J. Chem Phys. 79, 1487 (1983).

[15] M. Schlanges, Th. Bornath, D. Kremp, Phys. Rev. A88, 217 (1988).

[16] L. M. Biberman, V.S. Vorobjev, I.T. Yakubov, Kinetika Neravnovesnoi Niskotemperaturnoi Plasmy (Nauka, Moscow, 1982).

[17] H.W. Drawin, F. Emard, Physica 85C, 333 (1977).

TOKAMAKS

ATOMIC PROCESSES RELEVANT TO NEUTRAL BEAM BASED TOKAMAK DIAGNOSTICS

H.P. Summers, M. von Hellermann, P. Breger, J. Frieling, L.D. Horton,
R. Konig, W. Mandl, H Morsi, R. Wolf
JET Joint Undertaking, Abingdon, Oxon., OX14 3EA, U.K.

F.J. de Heer, R. Hoekstra
FOM-AMOLF, 1009 DB Amsterdam & KVI, 9747 AA Groningen, Netherlands

W. Fritsch
Hahn-Meitner Institute, Glienicker Strasse 100, Berlin, Germany

ABSTRACT

Charge exchange and beam emission spectroscopy in magnetic confinement fusion plasmas are reviewed. The range of spectral phenomenology is illustrated from the JET tokamak using fast deuterium and helium beams. The helium observations are new. The atomic reactions concerned in the emission and their interplays are summarised. The state of cross-section data for the analysis is briefly assessed.

INTRODUCTION

Neutral beam associated spectroscopic measurements have become firmly established over the last ten years as a major diagnostic tool for tokamak fusion plasmas. For fast penetrating beams such as $^1H^0$, $^2D^0$, $^3He^0$ and $^4He^0$ at energies ≥ 25 keV/amu, diagnostic advance has been very fruitfully associated with the implementation of such very powerful sources for supplementary heating of plasmas (1). Dedicated diagnostic beams of these types have since followed on. Proposals for the next international tokamak (ITER) include such H^0 and He^0 beams. By contrast, for the awkward environment of the plasma edge in large machines or for small diameter machines, the opportunities of using special beams such as $^7Li^0$ have long been realised (2).

Beam related spectroscopic measurements arise from two sets of reactions. Using a deuterium beam for illustration, these are

$$D^0_{beam}(1s) + X^{+z_0}_{plasma} \rightarrow D^+_{beam} + X^{+z_0-1}_{plasma}(nl) \qquad (1)$$

$$X^{+z_0-1}_{plasma}(nl) \rightarrow X^{+z_0-1}_{plasma}(n'l') + h\nu$$

in which charge exchange is the primary driving reaction and the emitted radiation gives information on the plasma ion $X^{+z_0}_{plasma}$. This is 'charge exchange spectroscopy' (CXS). The other set is

$$D^0_{beam}(1s) + \left\{ {X^{+z_0}_{plasma} \atop e} \right\} \rightarrow D^0_{beam}(nl) + \left\{ {X^{+z_0}_{plasma} \atop e} \right\} \qquad (2)$$

$$D^0_{beam}(nl) \rightarrow D^0_{beam}(n'l') + h\nu$$

in which beam excitation is the primary driving reaction and the emitted radiation characterises the beam atoms in the plasma environment. This is 'beam emission spectroscopy' (BES).

The advantages of such spectroscopy are :
   (a) Inducing of emission associated with non-radiating fully ionised impurity species in the plasma core. This is the main reason for the relevance of CXS in high temperature fusion plasmas. The induced radiation is diagnostic of ion temperature (3), plasma rotation (4) and impurity density. (5).
   (b) Localisation in space of the emission at the beam/viewing line intersection. For example, at JET a typical injector beam is ~ 300 cm² at the plasma centre. The optical

viewing line aperture is ~ 5 cm². For a 40 keV/amu $^2D^0$ beam, the radiative lifetime of the $n = 3$ shell is ~ 10^{-8} sec and implies a decay length along the beam line of ~ 2.5 cm. Therefore, for most purposes, the radiation is localisable and reflects the local reactions in the volume.

(c) Since the neutral beam atoms are traversing a magnetic field, the beam emission is in general resolvable into Zeeman, Paschen-Back **(2)** or motional Stark components depending on particle speeds and field strengths. **(6)**

(d) Discrimination of the beam/plasma emission from background and coincident radiation from the plasma edge is possible. Most precisely this is done by modulation of the beams and less well by pre-and post- beam measurement. In practice the difference in edge and central plasma feature widths often allows separation by constrained multi-gaussian fitting methods.

(e) Extended spectral regions over which CXS emission can be observed. This is especially true for fast beams because of the behaviour of charge exchange cross-sections to different quantum shells of the receiving ion and, in particular, allows easy visible wavelength observations.

The disadvantages are

(a) Attenuation of the beams. This necessitates careful calculation for quantitative studies. For example, in JET beam attenuation by a factor 100 can still gives measurable CXS spectral signals, but errors of ~ 20% in the stopping cross-section are amplified to errors ~ 100% in deduced impurity densities. It must also be noted that electron density measurements particularly near the edge of the plasma can be quite inaccurate.

(b) Mixed components in the beam. These may be fractional energy components in D^0 beams or metastable components in He^0 beams and add uncertainty to the analysis.

(c) The initially formed states following charge exchange from the donor beam species are disturbed by further collisions with plasma ions and electrons before radiating in general.

For this and the previous reasons quite elaborate theoretical calculations, using many cross-sections, are necessary to allow signal interpretation. There is another complication. The beam/plasma interaction establishes a number of secondary populations which in turn act as the sources for further reactions and there are additional radiating populations at the edge of the plasma. Both sets of populations can emit at the same wavelengths as the sought CXS lines. The secondary populations are of some importance. To summarise, there are :

(a) Halo atoms - These are D^0_{plasma} atoms formed by reactions (1) with $X \equiv D$. They migrate in a random walk by further CX reactions until ionised. Typically they are localised within ~ 30cm of the beam itself.

(b) Plume ions - These are ions formed by reactions (1) They travel along field lines. Excited by electrons they can radiate before finally ionising in positions some distance from their point of formation (in JET, ~ 40m for C^{+5} and ~ 6m for He^{+1} at 5keV & 5.0^{13} cm^{-3}) **(7)**.

(c) The slowing down ionised beam population. For example, for $^3He^0$ beams at 50 keV/amu in JET, the slowing down time of the fast $^3He^{+2}$ after double ionisation is ~ .3 - .6 sec.

(d) Ions $X^{+z_0 - 1}$ and $X'^{+z'_0 - 3}$ present at the periphery of the plasma. Note the spectral overlap of hydrogen-like and lithium-like emission.

It is to be noted that the population *(c)* is very similar to that of slowing alpha particles expected to be produced by D/T fusion **(8)**.

THE SPECTRAL PHENOMENOLOGY

In illustrating CXS and BES spectra from tokamaks, it must be noted that the particular geometric configuration of the machine and spectroscopic viewing lines influences strongly the relative importance of the various secondary populations. Equally the particular beam energies, concentrations of impurities, electron and plasma ion temperatures and densities determine the balance between the various atomic reactions. The data shown here are from the JET tokamak. The beam line configuration in JET is shown in plan view in figure 1, and has been described in detail elsewhere **(1)**.

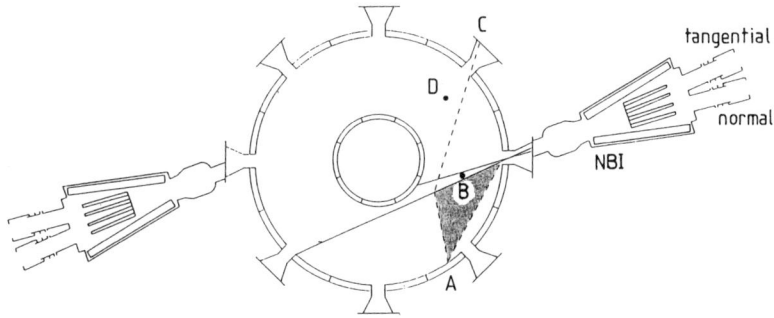

Fig. 1. Plan view of JET showing the two neutral beam injector assemblies organised as 'normal' and 'tangential' sets of four vertically arranged individual injectors. Spectral observations are allowed via a horizontal fan of viewing lines (A) ($\lambda \geq 4000\text{Å}$) and vertical lines at (B) ($\lambda \geq 3000\text{Å}$) and (D) ($\lambda \geq 4000\text{Å}$).

JET has been used to study beryllium as a limiter material and so provides a unique opportunity to observe beryllium CXS (9). Figure 2a shows the vicinity of 4685Å. The data was with deuterium beams at a primary energy of 40keV/amu. There are two CXS lines namely BeIV (n = 6-5) and BeIV (n = 8-6), distinctive because of their breadth. Each is a superposition of broad beam/plasma emission and edge emission. Figure 2b shows the same wavelength region but with modulated beams and appropriate subtractions.

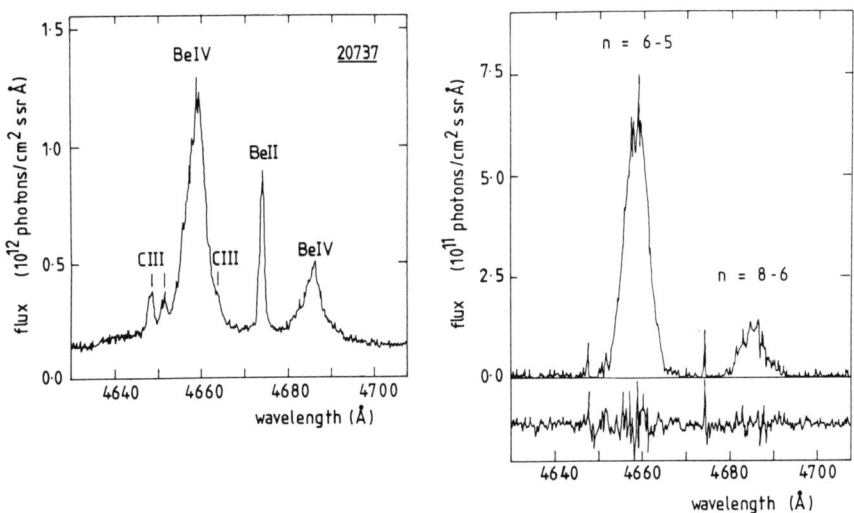

Fig. 2a & 2b. (a) Spectral observation through neutral D^0 beams showing edge (CIII, BeII and BeIV) emission and CXS (BeIV) emission. (b) The same spectral region with edge and background emission removed using modulated beam techniques. The remaining features are pure gaussians at the same (central plasma) temperature.

The edge emission and bremsstrahlung background are effectively eliminated leaving pure CX features. Their ratio therefore challenges atomic reaction modelling of the effective emission coefficients. In general other hydrogen-like or lithium-like species of even nuclear charge can emit at the same wavelength (for example, HeII (n = 4-3) or CVI (n = 12-9)). Helium was not present in the pulses shown and carbon was low. The edge emission in BeII 4673.5Å is a narrow feature at $T_i \sim 20$ eV and provides a wavelength fiducial. The centrally emitted charge exchange lines generally show an apparent plasma rotation. The rotation is real but its magnitude is modified by the energy dependence of the driving charge exchange reaction cross-sections. For the faster moving light plasma ions in particular, the excursion of their relative speed to the beam atoms from the beam speed produces spectral shifts, width and intensity alterations (10,11). These depend on the gradient and curvature of the reaction cross-section with relative speed. At $T_i = 20 keV$, $E_{beam} = 40 keV/amu$, $\Delta \lambda = 0.36 Å red$, $\Delta T_i = 6 keV$ for BeIV (n = 6-5). By contrast for the light ion He^{+2}, the temperature shift as observed in HeII ($n = 4-3$) at 150° to the beam line is $\Delta T_i = 9.6$ keV at $T_i = 20$ keV and $\Delta \lambda = 2.3 Å blue$.

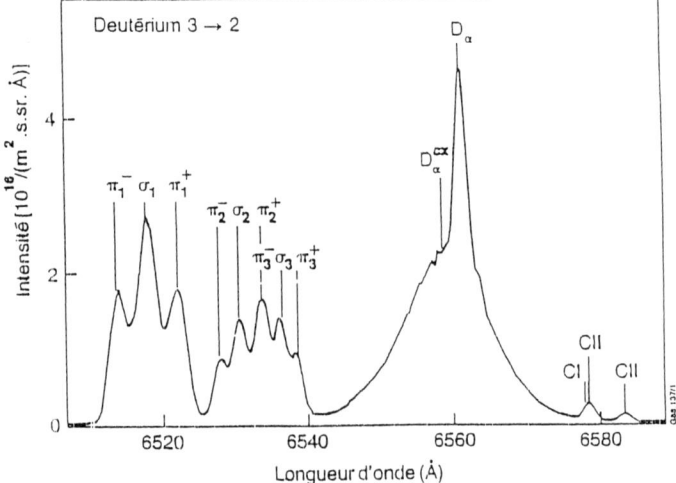

Fig. 3. Spectral observation through neutral D^0 beams in the vicinity of 6563Å showing edge D_α emission, unseparated CXS D_α emission and halo D_α emission, and full, half and third energy beam D_α emission. Note the doppler shift and Stark separation of the latter.

Figure 3 shows the spectral vicinity of D_α at 6563Å The beam emission is the composite feature on the blue side, doppler shifted because of the inclination of the beam line to viewing line. This is very convenient for distinguishing beam emission. The feature is composed of full, half and third energy parts with different mean doppler shifts. D_α emission from the beams is resolved into Stark multiplets because of the $\underline{v} x \underline{B}$ electric field in the frame of the moving beam atoms. These initially suprising features (6) are becoming familiar to tokamak diagnosticians. Figure 3 also identifies the various π and σ components. The information on internal magnetic field strength and orientation contained in these features is of great interest for tokamak studies. (12,13). On the other hand, the intensity of the features charts beam attenuation and collisionality through the plasma. Collisions with plasma and impurity ions are important here. Fluctuations in the beam intensity reflect density fluctuations in the plasma at least at frequencies accessible by the limits of signal integration time and on lengths scales set by radiative decay lengths (14).

Figure 4 shows again the spectral vicinity of 4685Å. However this experiment was with $^3He^0$ beams at 50 keV/amu. Helium was the dominant minority species in the plasma and the feature at 4685Å is HeII (n = 4-3) primarily. It is probably the first illustration of CXS with helium beams. Initial estimates of the effective emission coefficient for this CXS line with helium donor suggests that it is about 50% smaller than that with deuterium donor. On the other hand beam penetration is somewhat better so the overall impression is similar to that for deuterium beams. For other impurities such as beryllium and carbon, first impressions are that the CXS features are significantly weaker than for deuterium beams as indicated by our preliminary atomic cross-section data base for state-selective charge exchange capture with helium beams. Attention is drawn to the weak flat pedestal extending for 48 Å to the red side of 4685Å.

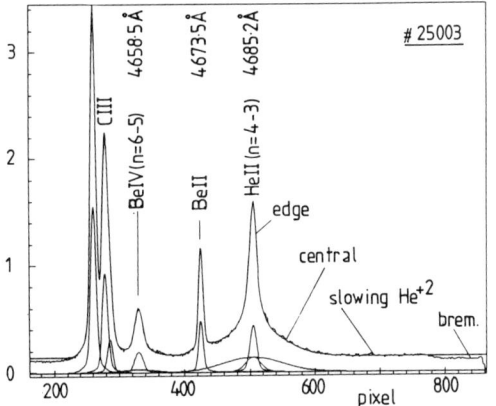

Fig. 4. Spectral observation through neutral $^3He^0$ beams showing edge (CIII, HeII, BeII and BeIV) emission, CXS (HeII) emission with the beams and emission from slowing He^{+2} (HeII).

This arises most probably from a fast $^3He^{+2}$ population in the plasma generated by the helium beams. The wavelength extension of the feature corresponds to a 50kev/amu doppler shift and the feature is assumed to extend similarly to the blue. It may be the first observation of a slowing alpha particle distribution.

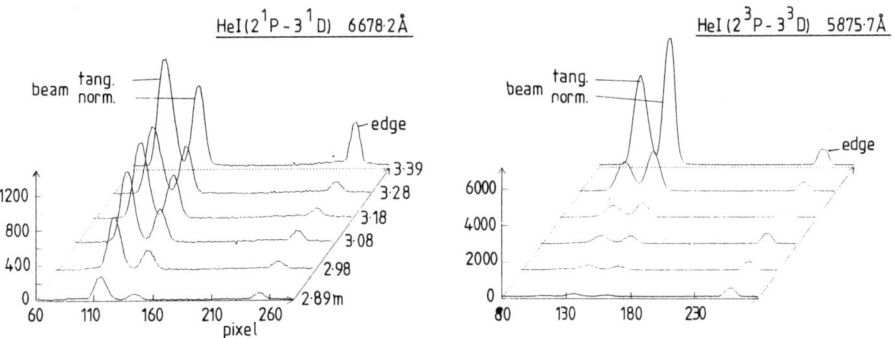

Fig. 5a & 5b. (a)Spectral observation through neutral $^3He^0$ beams in the vicinity of HeI (2p ^{1}P - 3d ^{1}D) at 6678 Å showing edge (HeI) emission and doppler shifted beam (HeI) emission. There are two beam lines contributing to the latter. (b) Spectral observation through neutral $^3He^0$ beams in the vicinity of HeI (2p ^{3}P - 3d ^{3}D) at 5876 Å showing edge (HeI) emission and doppler shifted beam (HeI) emission. There are two beam lines contributing to the latter.

Decay of the feature on time scales ~ 0.3 sec. after beam switch off indicates that the emission is not from detection of the helium beam itself as it ionises or by the helium beam as a CXS donor on the fast population but rather is associated with CX from thermal deuterium at the hydrogen rich edge 'X-point' regions of the plasma. In JET there is deuterium gas introduction at these points. This point is not yet fully demonstrated.

Finally figures 5a and 5b show the spectral vicinities of HeI ($2p^1P$ - $3d\ ^1D$) at 6678.2Å and of HeI ($2p^3P$ - $3d\ ^3D$) at 5875.7Å respectively. HeI edge emission is observed at the nominal wavelengths. The double feature in each case to shorter wavelengths is the helium beam emission. The viewing line intersects 'tangential' and 'normal' neutral beam lines with different inclinations. The strength of the triplet side spectrum in the beam is of note. The metastable content of helium beams formed by He^+/He neutralisation has been the subject of some discussion (15,16). It is expected to be < 7%. At a beam particle energy of 50keV/amu penetrating in a plasma of Z_{eff} = 1, the $2\ ^3S$ stopping length is ~ 50cm. compared with ~ 1m for $1\ ^1S$ at 5keV and 5.0^{13} cm^{-3}. The initial entrant metastable fraction, its destruction and the regeneration of the metastable population in the plasma is important for diagnostic studies. Evidently with such observations a complete CXS and perhaps BES is possible with helium beams as the donor.

ATOMIC REACTIONS AND CROSS-SECTIONS - HYDROGEN BEAMS

The atomic reactions of concern are firstly those responsible for beam stopping. At low density, these are direct losses from the beam atoms in their ground state. For $^2D^0$ beams it is given by the sum of the reactions (1) and the ionising reactions

$$D^0_{beam}(1s) + \left\{ \begin{matrix} X^{+z_0}_{plasma} \\ e \end{matrix} \right\} \rightarrow D^+ + \left\{ \begin{matrix} X^{+z_0}_{plasma} \\ e \end{matrix} \right\} + e \qquad (3)$$

Above 20 keV/amu ion collisions dominate electron collisions. In the large tokamaks, high ion temperatures, $T_i \leq$ 30keV, are encountered, so proper displaced Maxwell averages must be used in the equations. Cross-section data for D^+ and He^{+2} is good at all energies (10% accuracy at least is required). For the other light nuclei, the high energy region $E \geq$ 100keV/amu is good but ionisation cross-sections are poorly known below 80keV/amu. Although the stopping then is dominated by the charge exchange cross-sections, it is still difficult to be certain of 10% accuracy for ions of $z_0 \geq 3$

At high densities, stepwise losses via excited states become relevant, that is collisional-radiative rates are required. So excitation cross-sections matter. The excitation processes of course are those of equations (2) involved in formation of the beam emission spectrum. The situation for excitation cross-sections is much less satisfactory especially for $z \geq 2$. For ion collisions in the relevant regions 10 keV/amu $\leq E \leq$ 70 keV/amu, scaling rules of the $\frac{\sigma}{z} / \frac{E}{z}$ type (see (17) for a discussion) break down. Also little experimental data is available at these energies for $z_0 \geq 3$. This is the regime where the elaborate close-coupled atomic orbital calculations are of great assistance (18).

Figure 6 shows the new data for He^{+2}/H by Fritsch et al. (19) contrasted with older data. The oscillations are of note and influence modelling of higher density beam stopping and beam emission spectroscopy. Large scale population calculations making use of such data establish tabulations of beam stopping coefficients and effective emission coefficients of all spectrum lines of deuterium for all beam energies and plasma conditions. Figure 7 contrasts theory and experiment using multichord observations of D_α at JET. The primary impurity was carbon and Z_{eff} was ~ 3. Evidently this area of BES is progressing, justifying the cross-secion efforts.

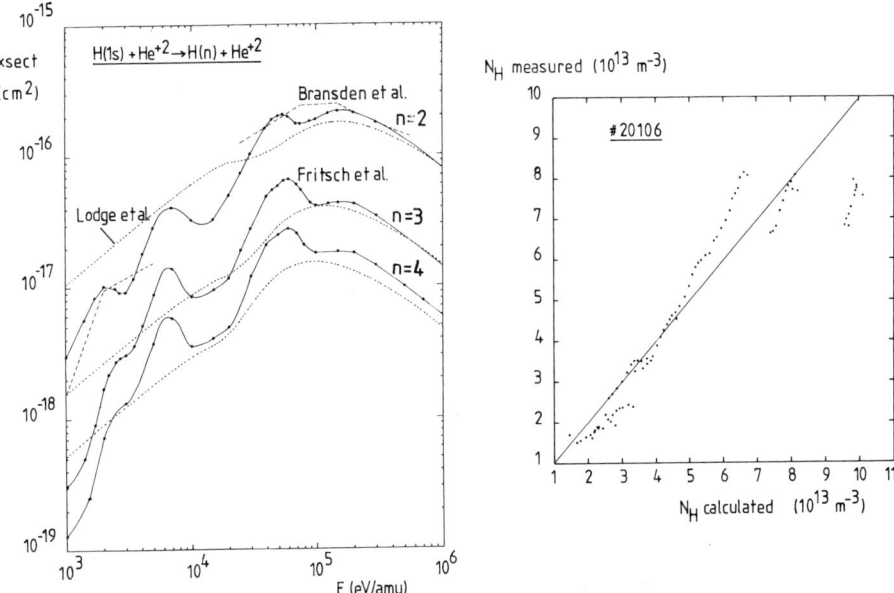

Fig. 6 & 7. (6) Comparison of $H(1s)$ excitation cross-sections due to He^{+2} impact. (7) Calculated and measured density of the primary energy fraction of the neutral deuterium beam in the plasma. The measured values are deduced from D_α.

CXS of impurities in tokamaks requires firstly state selective charge exchange cross-sections for the fully ionised ions of H, He and the first period atoms $Be - Ne$. Then, less routine, is CXS of heavier species such as iron, and also of partially stripped recombining ions particularly with helium-like or neon-like configurations. Since observations concentrate on the visible spectral region, capture cross-sections to n-shells with subdominant $n > z_0^{3/4}$ must be known and indeed the fractions into l-substates. Fields and ion collisions partially redistribute the l populations. This is very testing both for experiment and theory. The energy region 5 keV/amu to 200 keV/amu is required which spans from the molecular to the high energy regimes. Elaborate calculations and sophisticated experiments have contributed enormously to the available database in response to the fusion needs. In illustration, figure 8 reviews the state selective data for He^{+2}/H (11). Extensive collisional-radiative calculations of various types such as l-mixing cascade, bundle-nl and bundle-n (20) convert such data to the derived effective emission coefficients, mapped over plasma parameters, for the particular CXS lines used for experiment analysis. Improvements which take account of the secondary populations present further difficult demands on fundamental cross-section calculations. For example, the decreasing charge exchange capture to a high excited n-shell at lower energies means that CX from excited states of fractional energy beam atoms or from halo atoms have an opportunity to contribute. The role of excited state donors increases down to thermal edge features where it is completely dominant (21). Extensive calculations have been carried out extending the state selective CX database from excited donors over the whole collision energy range. Clearly such influences are hard to distinguish from structure in the ground state direct capture cross-sections shown in figure 8 and both must be accorded appropriate attention in the atomic modelling.

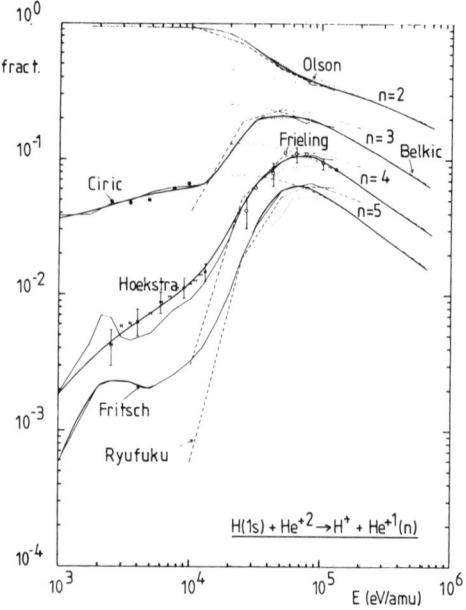

Fig. 8. New measurements and assessment of the partial capture cross-sections into n-shells in He^{+2}/H collisions.

ATOMIC REACTIONS AND CROSS-SECTIONS - HELIUM BEAMS

Helium beam stopping is more complicated than that of hydrogen. In addition to the single electron loss processes from the ground state as in equations (1) and (3), there are two electron loss processes of double ionisation, double charge transfer and transfer ionisation.

$$He^0_{beam}(1s^2\ ^1S) + X^{+z_0}_{plasma} \rightarrow He^{+2} + X^{+z_0}_{plasma} + e + e \qquad (4)$$

$$He^0_{beam}(1s^2\ ^1S) + X^{+z_0}_{plasma} \rightarrow He^{+2} + X^{+z_0-2}_{plasma}$$

$$He^0_{beam}(1s^2\ ^1S) + X^{+z_0}_{plasma} \rightarrow He^{+2} + X^{+z_0-1}_{plasma} + e$$

For He^{+2}, the double charge transfer is resonant. Figure 10 shows the various stopping cross-sections as used in the JET database for Be^{+4}. Note that at the 10% level of accuracy at 50 keV/amu all the processes need to be considered. For single CX there are no experiments, single ionisation below 100 keV/amu, transfer-double ionisation below 100keV/amu and double charge transfer are speculative. Comprehensive experimental data is available only for H^+ and He^{+2} and partial experimental data for Li^{+3} (22). At higher density and with a metastable content of the beam at source, the $2\ ^1S$ and $2\ ^3S$ metastables must be incorporated in the modelling of stopping since they are most influential on the stepwise collisional-radiative pathways. For BES, evidently transitions from the n = 3 to 2 levels on both the singlet and triplet side are the most immediately useful. The most interesting properties however are in the n = 4 shell. The $\underline{v \times B}$ motional Stark field mixes the $4f^1F$, $4d^1D$, $4p^1P$ progressively as the electric field increases up to 100 kV/cm. The resulting forbidden line positions are shown in figure 10. Unfortunately, unlike hydrogen the branching ratio $(1s^2\ ^1S - 1s4p\ ^1P)/(1s2s\ ^1S - 1s4p\ ^1P)$ is $\sim$ 34/1. With the mixing, transitions from the $4p$ to the ground act as a drain on the populations of the n = 4 levels and so the visible wavelength features must be very weak.

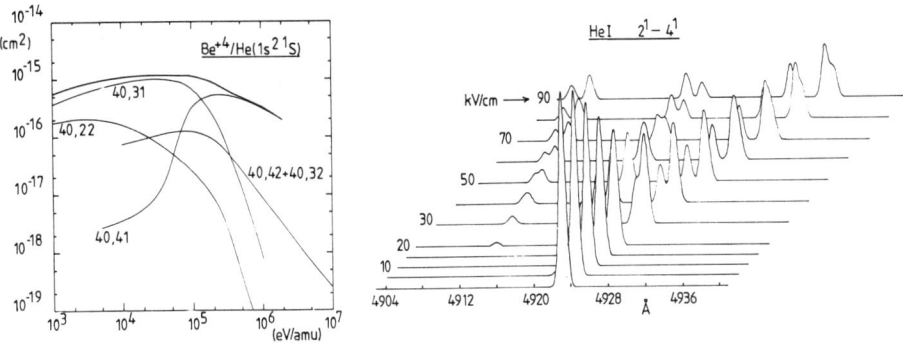

Fig. 9 & 10. (9) Stopping cross-sections for $He(1\ ^1S)$ in collision with Be^{+4} at zero density. (10) Theoretical Stark patterns for HeI ($n=2$-4) singlets with increasing electric field strength viewed orthogonal to the electric field.

These features have not yet been observed on JET at integration times of 1 sec. although blue enhancement of detectors may allow this at a later stage.

Population of the triplet side in the plasma from the ground state is difficult, since spin changing collision cross-sections by electrons become very small at $T_e \geq 2$ keV. Population is more likely through spin system breakdown of 4 ^1F and 4 ^3F under the Breit interaction. Clearly more observations of helium beams are required before the helium BES is fully understood and the opportunities for diagnostic application explored. Turning to CXS with helium beams, the data base of state selective charge exchange cross-sections to subdominant n-shells in the 10 keV/amu - 100 keV/amu range appears to be almost non-existent except with H^+ (see Hoekstra et al. - this meeting) although the situation is better at higher energies. A transfer database from deuterium is at present in use at JET, based on overbarrier model considerations. Its validity remains uncertain although initial comparisons with observations are encouraging. The role of the helium metastables as donors remains ambiguous. For $C^{+5}(n=8)$ estimates with JET beams suggest that the 2 ^3S CX contribution is less than that of 1 ^1S. Evidently considerable effort is required to bring CXS with helium beams to the same state as that with hydrogen beams. It is likely that helium beams in the 30 keV/amu - 50 kev/amu will be used quite extensively in fusion plasma studies and so may justify such effort.

ATOMIC REACTIONS AND CROSS-SECTIONS - LITHIUM BEAMS

Use of low energy lithium beams, $E \sim 10$ keV/amu parallels that of the hydrogen and helium fast beams. The short ionisation lengths of Li^0 in a plasma (~ 10 cm decay length for 60 keV beams at density $\sim 2.0^{13}$ cm^{-3}) makes it most valuable as an edge probe. The decay of the resonance line radiation at 6707.8Å is described with multi-level ($n \leq 3$ typically) collisional-radiative models.

Measurements along the decay path of the beam allow reconstruction of the density profile (23). Electron and ion collisions are both involved. Likewise CXS with Li^0 as the donor has been carried out (for example on C^{+6} and C^{+5} as receiver (24)). After spectral separation of the π component of the resonance line (Pashen-Back separations of π and σ components are typically $\sim$ 0.2Å) measurement of its polarisation angle has been successfully used to obtain the internal magnetic field orientation and hence the current distribution in small tokamaks (2).

CONCLUSIONS

The paper has sought to show some of the highlights of charge exchange spectroscopy and beam emission spectroscopy in tokamak plasmas. Extensive modelling of the collisional-radiative type is required for full exploitation of these active diagnostics. The enormous international effort on fundamental ion/atom collision cross-sections has given credibility to this, such that increasingly precise and subtle deductions are pursued. All modern large tokamak enterprises now depend on their CXS and BES diagnostics.

REFERENCES

(1) A Boileau, M von Hellermann, L D Horton & H P Summers (1989) Plasma Phys.and Contr. Fusion 31, 779.
(2) K McCormick (1986) Proceedings of 'Basic and Advanced Diagnostic Techniques for Fusion Plasmas', Varenna, p635.
(3) R J Fonck, R J Goldston, R Kaita & D E Post (1983) Appl. Phys. Lett. 42, 239.
(4) R J Groebner, N H Brooks, K H Burrell & L Rottler (1983) Appl. Phys. Lett. 43, 920.
(5) R C Isler (1981) Phys. Rev. A 24, 2701.
(6) A Boileau, M von Hellermann, W Mandl, H P Summers, H Weisen & A Zinoviev (1989) J. Phys. B 22, L145.
(7) R J Fonck, D S Darrow & K P Jaehnig (1984) Phys. Rev. A 29, 3288.
(8) D E Post (1981) J. Fusion Energy 1, 129.
(9) H P Summers, W J Dickson, A Boileau, P G Burke, B Denne-Hinnov, W Fritsch, R Giannella, N C Hawkes, M von Hellermann, W Mandl, N J Peacock, R Reid, M F Stamp & P R Thomas (1991) Plasma Phys.and Contr. Fusion - in press.
(10) M von Hellermann, W Mandl, H P Summers, H Weisen, A Boileau, P D Morgan, H Morsi, H Koenig, M F Stamp & R Wolf (1990) Re. Sci. Intrum. 61, 3479.
(11) M von Hellermann, W Mandl, H P Summers, A Boileau, R Hoekstra, F J de Heer & J Frieling (1991) Plasma Phys. and Contr. Fusion - in press.
(12) C Challis, M von Hellermann, B Keegan, R Konig, W Mandl, J O'Rourke, R Wolf & W Zwingmann (1991) '18th European Physical Society Conf. on Controlled Fusion & Plasma Physics, Berlin. JET-P(91)08.
(13) D Wroblewski, K H Burrell, L L Lao, P Politzer & W P West (1990) Rev. Sci. Instrum. 61,3552.
(14) S F Paul & R J Fonck (1990) Rev. Sci. Instrum. 61,3496.
(15) K Tobita, T Itoh, A Sakasai, Y Kusama, Y Ohara, Y Tsukahara, M Nemoto, Y Kawano, H Kubo, H Takeuchi & T Sugie (1990) Plasma Phys. and Contr. Fusion 32, 429.
(16) R S Hemsworth (1991) - private communication
(17) W Fritsch & K-H Schartner (1987) Physics Letters 126, 17.
(18) W Fritsch & C D Lin (1991) Physics Reports 202, 1.
(19) W Fritsch, R Shingal & C D Lin (1991) Phys. Rev. A - in press.
(20) J Spence & H P Summers (1986) J. Phys. B 19, 3749.
(21) M Mattioli, N J Peacock, H P Summers, B Denne & N C Hawkes (1989) Phys. Rev. A 40, 3886.
(22) M B Shah & H B Gilbody (1985) J. Phys. B 18, 899.
(23) F Aumayr, M Schneider, E Unterreiter & H. Winter (1991) - .
(24) R P Schorn, E Hintz, D Rusbuldt, F Aumayr, M Schneider, E Unterreiter & H Winter (1991) J. Phys. B 91, - .

POLARIZATION SPECTROSCOPY OF TOKAMAK PLASMAS

Dariusz Wróblewski
Lawrence Livermore National Laboratory, Livermore, CA 94550

ABSTRACT

Measurements of polarization of spectral lines emitted by tokamak plasmas provide information about the plasma internal magnetic field and the current density profile. The methods of polarization spectroscopy, as applied to the tokamak diagnostic, are reviewed with emphasis on the polarimetry of motional Stark effect in emissions of hydrogenic neutral beams.

INTRODUCTION

High temperature plasma is confined in a tokamak by an externally produced toroidal magnetic field (long way around the torus) and a poloidal magnetic field (short way around the torus) associated with the toroidal current induced in the discharge (Fig. 1). The resulting helical field forms a set of nested magnetic surfaces. The stability and confinement properties of the plasma depend critically on the distribution of the plasma current density. The current profile may be considered to be one of the most important parameters of the tokamak discharge but it is also one of the most difficult to measure.

In the usual approach, the current distribution is inferred from a measurement of the poloidal magnetic field profile. As the poloidal magnetic field is usually much smaller than the externally produced toroidal field, the measurement error smaller than 1% of the total field, with the spatial resolution of the order of 1 cm, is required to provide a meaningful estimate of the current profile. The time resolution of the order of 10^{-3} sec has been achieved by some of the diagnostics but an extension of this range to 10^{-5}–10^{-6} sec will be required for the studies of plasma fluctuations and details of magnetohydrodynamic (MHD) instabilities.

With the growing importance of the current profile measurement in the tokamak, a number of measurement techniques has been proposed and implemented. This report concentrates on the polarization spectroscopy methods which take advantage of the polarization properties of spectral lines emitted by a magnetized plasma. The polarization may be due to the Zeeman effect or, in the case of lines emitted by high energy hydrogenic beam particles, to the Stark effect. In particular, the dependence of the polarization on the direction of the local magnetic field and the observation direction is used to determine the direction of magnetic field. For plasmas with circular cross-section and when the modification of the externally produced toroidal field by the plasma is negligible, the current profile may be easily

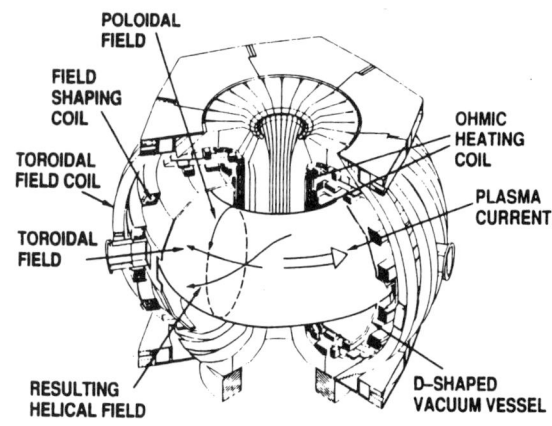

Fig. 1. View of GA's DIII-D tokamak showing the magnetic field coils and the directions of magnetic field components.

© 1992 American Institute of Physics

deduced from the polarization measurement. In the case of shaped plasmas with high stored energy density, the MHD equilibrium constraint is invoked to reconstruct the current profile from the magnetic field measurements.

The success of the polarization spectroscopy in the determination of the plasma current profile depends on the presence of a suitable source of line radiation in the plasma and on a very accurate measurement of its polarization. With the Doppler broadening dominating the line profiles, the net polarization of the most prominent spectral features of a tokamak plasma, located in the UV and XUV parts of the spectrum, is negligible. However, as the ratio of the Zeeman splitting to the Doppler broadening increases with the transition wavelength, the polarization fraction of the visible and near UV lines may be measurable. Also, the instrumentation capable of measuring the very small polarization effects is not available for wavelengths below the UV range. Thus, only the transitions in near-UV or visible range of wavelengths may be employed for the polarization measurements. The types of emissions that have been considered include the spectral lines emitted by the neutral particles injected into the plasma as a high energy neutral beam[1-5] or a pellet,[6,7] the forbidden transitions in heavy impurity ion lines,[8-10] and the charge-exchange recombination lines.[11] The hydrogenic emissions from high energy beams, with their spectrum dominated by strong motional Stark effect (MSE), proved to be especially useful for the current profile diagnostic.[12-14]

POLARIMETRY OF ZEEMAN EFFECT

In the weak field approximation, the wavelength separation (in Å) of the Zeeman components of a spectral line is given by (cf. e.g. Ref. 15):

$$\Delta\lambda_B = 4.67 \times 10^{-9} z B \lambda_0^2 \Delta M, \qquad (1)$$

where B is the total magnetic field (in Tesla), λ_0 is the line wavelength (in Å), and z is the effective splitting factor (intensity weighted average of splitting factors of a blended group of transitions with the same polarization properties[16]). There are two types of transitions: π ($\Delta M = 0$) which are linearly polarized, and σ ($\Delta M = \pm 1$) which, in general, are elliptically polarized. For the observation along the magnetic field direction only the σ-components are observable and are, respectively, lefthand or righthand circularly polarized. When observed from a direction perpendicular to the magnetic field the radiation is linearly polarized either parallel to the magnetic field ($\Delta M = 0$) or perpendicular to the magnetic field ($\Delta M = \pm 1$). [The above rules are for the electric dipole transitions. For the magnetic dipole (forbidden) transition the directions of linear polarization vectors should be reversed].

The first diagnostics of the poloidal magnetic field were based on the measurement of the polarization direction of the Zeeman π-component. In this case, the observation direction is perpendicular to the total magnetic field (i.e. along the tokamak major radius) and the change of the polarization direction due to the plasma current gives the magnetic field pitch angle $\gamma = \mathrm{atan}(B_p/B_T)$, where B_T and B_p is the toroidal and the poloidal component of the magnetic field, respectively.

The linear polarization of resonance line of high energy ($\sim$60 keV) lithium neutral beam was measured in the experiments on ASDEX[1,2] and TEXT[3,4] tokamaks. The lithium beam injected into the tokamak plasma produces a strong resonance line with small Doppler broadening and, because of its low atomic number, causes a negligible perturbation of the plasma. Also, a very good spatial resolution may be achieved if a small diameter beam is used. About 0.1 degree accuracy of the tilt angle measurement was achieved with 50 ms integration time.[2] Unfortunately, because of the large ionization cross-section and poor penetration, the usefulness of the lithium beam is limited to low density, small size plasmas. In order to avoid the attenuation problem, Levinton proposed to use a helium beam which has much lower ionization cross-section than lithium.[5]

As a low cost alternative to high energy particle beams, a high velocity pellet may be employed. As the pellet travels through the plasma its material is evaporated and excited, and the polarization of spectral lines may be analyzed to yield information about the local magnetic field. Lithium pellets were employed in the experiments on the Alcator C and TFTR tokamaks[6,7] and polarization measurements were obtained for the LiII lines ($\lambda \simeq$ 5484 Å). The pellet spectroscopy provides a "snapshot" of the poloidal field profile as the pellet transverses the plasma in few hundred microseconds, followed by a large perturbation of the plasma due to the deposition of the pellet material in the plasma interior. Thus, this approach is not useful for continuous monitoring of the plasma current profile.

As the spectral separation of the circularly polarized Zeeman components is twice as large as separation of the linearly polarized components, the measurement of the fractional circular polarization in the lines for which $\Delta\lambda_B/\Delta\lambda_D \leq 1$, where $\Delta\lambda_D$ is the Doppler broadening, may result in better sensitivity than the measurement of the linear polarization. This approach takes advantage of the fact that the polarization of σ-components varies from purely linear to purely circular when the direction of observation is changed from perpendicular to parallel to the magnetic field direction. Thus, the relative content of circular polarization is a direct measure of the magnetic field component in the direction of observation. The measured quantity is the difference between the righthand and the lefthand circularly polarized line profiles:

$$I_L - I_R = \cos\gamma_0 [I(\lambda + \Delta\lambda_B) - I(\lambda - \Delta\lambda_B)], \qquad (2)$$

where γ_0 is the angle between the direction of observation and the total magnetic field, λ is the wavelength, and $I(\lambda)$ is the function describing the polarized line profile (a Gaussian for high temperature plasma and/or for lines with simple Zeeman patterns). For $\Delta\lambda_B/\Delta\lambda_D \ll 1$:

$$\max(I_L - I_R) \sim \cos\gamma_0 \frac{\Delta\lambda_B}{\Delta\lambda_D} \sim B\cos\gamma_0 \left(\frac{A}{T_i}\right)^{1/2}, \qquad (3)$$

where T_i is the ion temperature, and the maximum of $(I_L - I_R)$ is at $\lambda - \lambda_0 \approx \Delta\lambda_D/2$. Thus, $\max(I_L - I_R)$ is directly proportional to the component of the magnetic field in the direction of observation. $\Delta\lambda_D$ and $\Delta\lambda_B\cos\gamma_0$ may be determined by fitting the measured $I_L - I_R$ and line profiles, yielding the ion temperature and the component of magnetic field in the direction of observation, respectively.[17] For the poloidal field measurement in a tokamak, the observation must be in the major radial direction (to avoid contribution from the toroidal field) and tangent to the magnetic surface on which the field is to be measured. In this case, $\cos\gamma_0$ will vary from zero in the plasma center to about 0.1 at the plasma edge.

Feldman et al.[8] proposed to measure the circular polarization of the forbidden (magnetic dipole) lines produced by transitions within ground configuration of high ionization stages of heavy impurity ions (Fe, Ti, Cr, etc.). As the ratio $\Delta\lambda_B/\Delta\lambda_D$ increases with wavelength and atomic weight the UV/visible forbidden lines have relatively large fractional polarization. No injection is necessary if an intrinsic heavy impurity is present. A successful measurement of the forbidden line TiXVII 3834 Å (transition $2s^2 2p^2\ ^3P_2 \rightarrow\ ^3P_1$) was obtained on the TEXT tokamak (University of Texas, Austin). A single sightline, scanning polarimeter was employed to measure both the line and $I_L - I_R$ profiles. Intrinsic titanium originated from the machine limiter. The line showed a substantial brightness throughout the plasma cross section and the whole poloidal field profile could be obtained on shot-to-shot basis. The Abel inversion was used to obtain the local quantities. Figure 2 shows an example of the poloidal field profile measured for a discharge with $B_T = 2.0$ T, plasma current I = 200 kA, and plasma density 10^{-13} cm^{-3}. Each of the data points represents an average of the results obtained from 5 to 10 single measurements. Accuracy of the fractional circular polarization measurement was of the order of 5×10^{-3} with the integration time of the order of 20 ms. About 0.005 T accuracy of the poloidal field measurement was achieved with 50 ms integration time and superposition of 5–10 single measurements.[10,17]

Similar approach was used to measure the circular polarization of resonance line of the lithium beam.[18] Comparable accuracy of the diagnostic was achieved with the additional advantage of local instead of line integrated measurement.

A measurement of circular polarization of light impurity lines (He, C) was attempted in the DIII–D tokamak. The hydrogenic transitions produced as a result of charge exchange between fully ionized light impurities and high energy neutral beam particles were studied. Even at high plasma temperatures the charge-exchange recombination lines should exhibit a measurable circular polarization fraction (10^{-2} to 10^{-3}). In the DIII–D experiments, the charge-exchange lines were found to be overwhelmed by their strong, collisionally excited counterparts emitted from the plasma edge which led to inadequate accuracy of the polarization measurement.[19]

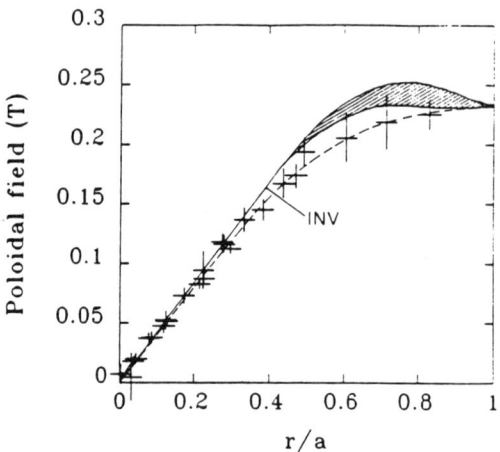

Fig. 2. Poloidal field profile in the TEXT tokamak deduced from the circular polarization measurement of the titanium impurity line. Dashed line is a fit to the measured profile and solid line (INV) shows the result of Abel inversion. Here, a is the plasma minor radius and r/a is the normalized radius.

MOTIONAL STARK EFFECT

Recently, very encouraging results have been obtained with the use of neutral beam emissions of hydrogenic species.[12,14] In this case, the polarization is due the Stark effect in the Lorentz ("motional") electric field $\mathbf{E} = \mathbf{v} \times \mathbf{B}$, where $\mathbf{v}$ is the velocity of the neutral particle, and $\mathbf{B}$ is the tokamak magnetic field. The linear Stark effect is observed in the hydrogenic species (hydrogen or deuterium) leading to large splitting and substantial fractional polarization of beam emitted lines. (The line broadening is due almost entirely to the beam divergence). The Doppler shifted emissions are observed and thus the interference of the strong line emissions from the plasma edge is avoided. Either a high power heating neutral beam or a dedicated low power "diagnostic" beam may be used. The spatial resolution is determined by the beam size and the measurement geometry. The Stark effect produces linear polarization only and, similar to the measurements of Zeeman effect linear polarization, direction of the local magnetic field is determined by measuring the direction of linear polarization of the emission.

Figure 3 shows a spectrum of the Balmer-α line emitted by a 75 keV hydrogen beam as observed by a tangentially viewing spectrometer in DIII–D tokamak. The Doppler shifted beam emissions are observed at full, 1/2 and 1/3 beam energy, corresponding to the molecular composition of the beam.[20] For a given energy component, three spectral features are resolved which are identified as clusters of Stark σ and π-components.[20,21] The σ-components are linearly polarized in the direction perpendicular to the electric field direction when observed in the direction perpendicular to the electric field and are not polarized when observed along the field. The π-components are polarized parallel to the electric field when observed in the direction perpendicular to the field and have null intensity for observation along the field. The magnetic field pitch angle is deduced from a measurement of direction of polarization of the full energy σ-components.

MODULATION TECHNIQUES IN POLARIZATION MEASUREMENTS

The polarization measurements require an efficient elimination of the unpolarized background and an accurate relative calibration of measurements of different polarization states. The modulation techniques accomplish that by converting the polarization content information into intensity modulation that may be accurately measured using the phase-sensitive (lock-in) detection. This approach has been widely used in the astronomical applications[22] and recently, in the polarimetry of tokamak plasmas.[10–14,17,18]

Figure 4 shows a schematic of modulated analyzer of linear polarization. The system consists of two photoelastic modulators (PEM$_1$ and PEM$_2$) and the linear polarizer (or beam splitting polarizer). The photoelastic modulator[23] is a fused silica plate in which a time-dependent birefringence is produced by a resonant, standing compression wave. The device offers large usable aperture and very large acceptance angle. For the linear polarization measurements, the modulators act as oscillating half-wave plates, i.e. the retardation is varied between approximately $+\lambda_0/2$ and $-\lambda_0/2$, at frequencies ω_1 and ω_2 (e.g. 42 and 47 kHz for the present DIII-D system), leading to a modulation in the direction of linear polarization of the transmitted radiation. The conversion to the intensity modulation is accomplished by the linear polarizer. With a proper orientation of the polarizing elements (axes of the modulators at -22.5 and $+22.5$ degrees with respect to polarizer axis) the amplitudes of intensity modulation at the second harmonics of the modulators frequencies are given by:

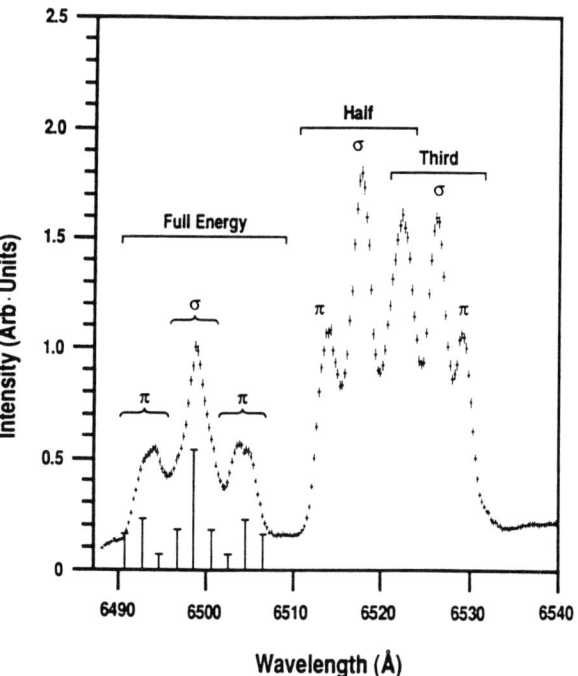

Fig. 3. Spectrum of the Balmer-α line emitted by hydrogen neutral beam in DIII-D. Observation geometry as shown in Fig. 5, $R = 2.14$ m.

$$I_\pm(2\omega_1) = \pm \frac{I_\sigma - I_\pi}{\sqrt{2}} J_2(\phi_1) sin 2\gamma_m, \qquad (4)$$

$$I_\pm(2\omega_2) = \pm \frac{I_\sigma - I_\pi}{\sqrt{2}} J_2(\phi_2) cos 2\gamma_m, \qquad (5)$$

where: $I_\sigma(\lambda)$ and $I_\pi(\lambda)$ are the intensities of Stark σ and π features, respectively, J_2 is the Bessel function of the first kind, ϕ_1, ϕ_2 are the retardation maxima ($\simeq 3.1 (\simeq \lambda/2)$), and γ_m is the measured magnetic field tilt angle. The "$\pm$" sign indicates that I_- and

I_+ signals are out of phase. Thus, the magnetic field tilt angle is obtained as a ratio of modulation amplitudes detected at frequencies $2\omega_1$ and $2\omega_2$:

$$tg2\gamma_m = \frac{I_+(2\omega_1) - I_-(2\omega_1)}{I_+(2\omega_2) - I_-(2\omega_2)} = \frac{I_+(2\omega_1)}{I_+(2\omega_2)}. \tag{6}$$

The modulation amplitudes are measured by two lock-in amplifiers referenced to the resonant frequencies of the modulators. Note that a single detector may be used to measure the signals at both modulation frequencies and that only the polarized light (with intensity $\sim (I_\sigma - I_\pi)$) contributes to the signal. In order to increase the number of collected photons, two detectors are used to measure both I_- and I_+ signals, which are later combined in the detection circuit.

MSE POLARIMETRY IN DIII–D

The measurement geometry used by the DIII–D polarimeter is shown in Fig. 5. Up to 8 high power neutral beams (hydrogen or deuterium) are injected on the machine midplane at an angle oblique to the toroidal magnetic field. The radiation is collected from a 2–6 cm diameter volume at the intersection of one of the neutral beams and the polarimeter sightline. The sightline may be moved within the region indicated approximately on the figure to provide a shot-to-shot profile measurement. In the absence of plasma current the motional electric field is vertical. By measuring the change of the polarization direction of the Stark σ-components when the plasma current is introduced, the magnetic field pitch angle may be determined.

In the polarimeter employed on the DIII–D tokamak the polarization analyzer is placed on the tokamak viewing port and the modulated light is transmitted by a fiber optic link to the interference filter spectrometer. There, a narrow bandpass (3 Å) interference filter is used to select the wavelength of the Doppler shifted σ-components. The peak transmission wavelength of the filter may be tuned over a limited range (to match the Doppler shift) by changing the incidence angle. Two photomultiplier tubes are used to detect the I_- and I_+ intensities.[14]

An example of high quality data that may be obtained from the MSE measurement is demonstrated in Fig. 6

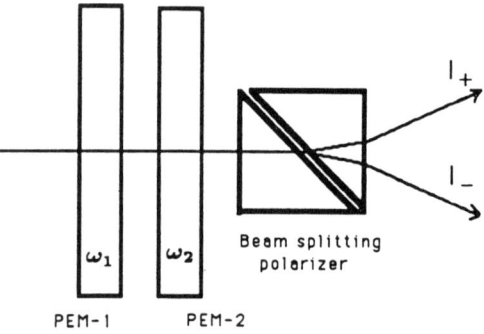

Fig. 4. Schematic of the linear polarization analyzer employing the photoelastic modulators.

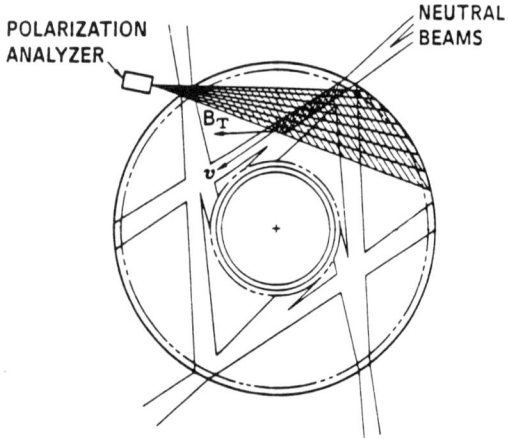

Fig. 5. Top view of the DIII–D tokamak showing the location and spatial coverage of the MSE diagnostic. Tokamak major radius $R_0 = 1.67$ m, the polarimeter covers the range $R = 1.5$–2.3 m.

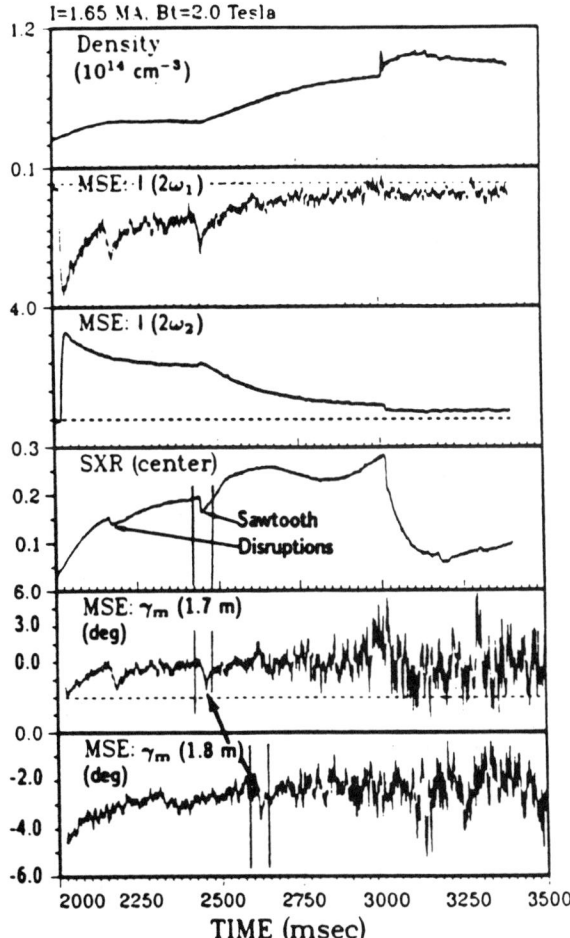

Fig. 6. Time history of a tokamak discharge showing the line integrated density, polarimeter signals $(I(\omega_1) \sim \sin\gamma_m$, and $I(\omega_2) \sim \cos\gamma_m)$, soft x-ray emission, and the magnetic field tilt angle measured at $R = 1.7$ m and $R = 1.8$ m. The timeslices used for the analysis in Fig. 7 are indicated by the vertical lines. The last trace is from a different but similar discharge, where the analyzed sawtooth collapse occurs somewhat later.

which shows a fragment of the time history of 1.65 MA, 2.0 Tesla beam heated discharge. In the time interval shown the plasma density increases leading to a decrease in the neutral beam penetration and deterioration of the polarimeter signals. Nevertheless, an accurate measurement of the tilt angle in the plasma center is obtained for the line average plasma density up to 6×10^{13} cm^{-3}. The trace labeled SXR shows the intensity of soft x-ray emissions as observed by a sightline close to the plasma center. This trace indicates a presence of the "sawtooth" instability, manifested as plasma disruption in the central part of the discharge, leading to a flattening of the density and temperature profiles, and redistribution of the plasma energy from the center to the outside. Note the correlation between the SXR signals and the measured magnetic field tilt angle indicating that a change in the current profile is associated with the instability.

For the highly shaped DIII–D plasmas, the current profile cannot be deduced from the internal magnetic field measurements only. The MSE data is incorporated into the magnetic field equilibrium code EFIT.[24] The code uses the data from an array of external magnetic measurements and, by invoking the force balance equation and the toroidal symmetry, calculates the global plasma parameters (current, energy content, position and shape of the last closed surface, etc.). The incorporation of the internal magnetic field measurements allows us to calculate also the shape of internal magnetic surfaces and the current profile. Optionally, the measured plasma pressure profile or the soft x–ray emission profile may be used in the equilibrium calculation to provide a reconstruction of the current density profile consistent with all available diagnostics.

Figure 7 shows an example of the equilibrium reconstruction obtained with the use of shot-to-shot MSE profile measurement. External magnetic data from six discharges similar to the one shown in Fig. 6 were averaged for the timeslices before and after the sawtooth disruption. The figure shows a poloidal cross-section of the tokamak with the reconstructed flux surfaces and the positions of the MSE measurements indicated by "+". The current profile obtained from this analysis shows the expected flattening in the center of the discharge as a result of the sawtooth disruption.

SUMMARY AND CONCLUSIONS

Rapid progress in the accuracy and reliability of the polarization spectroscopy diagnostics has been achieved in recent years. The spectroscopic techniques are based on well understood physical principles and share the advantage of employing visible lines, for

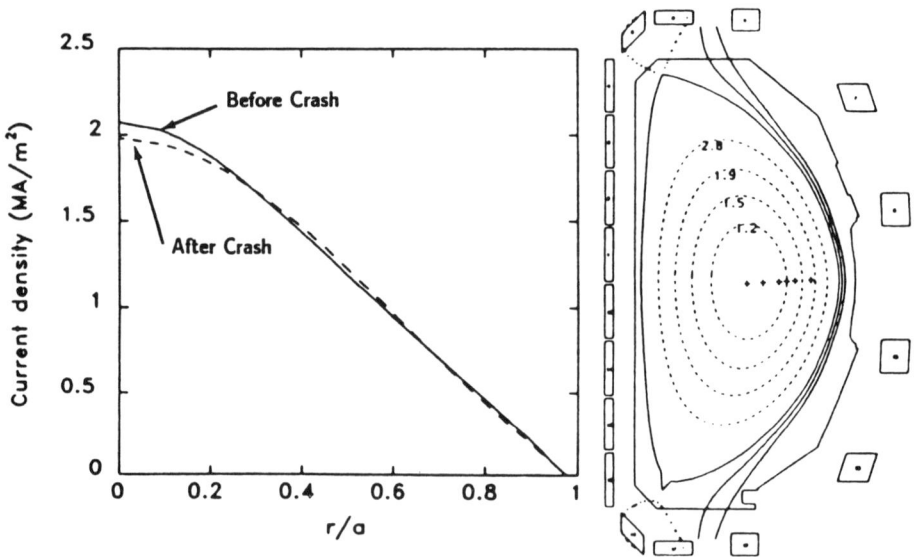

Fig. 7. The profile of flux surface averaged current density and the magnetic surfaces obtained from the equilibrium reconstruction using external magnetic field measurements and a 6 point MSE scan.

which easily available, relatively simple to operate and inexpensive instrumentation may be used. The modulation techniques, used before in similar astronomical applications, were found to be extremely useful for the tokamak diagnostic.

The presence of a suitable source of polarized radiation in the plasma interior is one of the prerequisites for a successful diagnostic. Although a number of results important for the tokamak physics has been obtained with the use of high energy lithium beams, pellets, and heavy impurity ion lines, the usefulness of these emissions seems to be limited to particular plasma conditions (low density, large magnetic field, etc.). Presently, the polarimetry of the motional Stark effect appears to be the diagnostic of choice, specially for large size tokamaks. Very good accuracy of the measurement was demonstrated with prototype instruments[12,14,25] (e.g. about 0.1 degree accuracy in the tilt angle measurement with 10 ms integration time in DIII-D). The major advantages of the beam emitted hydrogenic lines are their large partial polarization, reduction of thermal background by observation of Doppler shifted emissions, and generally strong emissivities from the plasma center. As the high energy neutral beams are routinely employed for plasma heating, the source of radiation is in most cases readily available although large size of a heating beam may lead to deterioration of the spatial resolution of the diagnostic. With the advance of the motional Stark effect approach the measurement of tokamak current profile, not long ago thought to be almost impossible, has been demonstrated as a routine diagnostic and has provided valuable insights into the physics of tokamak plasmas.

The time resolution obtained to date is still not sufficient for the studies of plasma fluctuations and details of MHD instabilities. The presently used modulation techniques limit the time resolution to about 0.1 ms (much longer than the time scale of the modulation cycle). The polarization measurement accuracy is ultimately limited by the photon flux available for analysis and some improvement may be obtained by employing more efficient collection and polarization optics.

ACKNOWLEDGMENTS

I would like to thank the DIII-D Team for their hospitality and support during my assignment at General Atomics.

Work performed under the auspices of the U.S. Department of Energy by the Lawrence Livermore National Laboratory under Contract No. W-7405-ENG-48.

REFERENCES

1. K. McCormick and J. Olivain, Rev. Phys. Appl. **13**, 85 (1978).
2. K. McCormick, et al., Phys. Rev. Lett. **58**, 491 (1987).
3. W. P. West, Rev. Sci. Instrum. **57**, 2006 (1986).
4. W. P. West, D. M. Thomas, J. S. deGrassie, and S. B. Zheng, Phys. Rev. Lett. **58**, 2758 (1987).
5. F. M. Levinton, Rev. Sci. Instrum. **57**, 1834 (1986).
6. E. S. Marmar, J. L. Terry, B. Lipschultz, and J. E. Rice, Rev. Sci. Instrum. **60**, 3739 (1989).
7. J. L. Terry, E. S. Marmar, R. B. Howell, et al., Rev. Sci. Instrum. **61**, 2908 (1990).
8. U. Feldman, J. F. Seely, N. R. Sheeley, Jr., S. Suckewer, and A. M. Title, J. Appl. Phys. **56**, 2512 (1984).
9. D. Wróblewski, H. W. Moos, and W. L. Rowan, Appl. Phys. Lett. **48**, 21 (1986).
10. D. Wróblewski, L. K. Huang, H. W. Moos, and P. E. Phillips, Phys. Rev. Lett. **61**, 1724 (1988).
11. D. Wróblewski and H. W. Moos, Rev. Sci. Instrum. **57**, 2029 (1986).
12. F. M. Levinton, R. J. Fonck, G. M. Gammel, R. Kaita, H. W. Kugel, E. T. Powell, and D. W. Roberts, Phys. Rev. Lett. **63**, 2060 (1989).
13. F. Levinton, G. M. Gammel, R. Kaita, H. W. Kugel, and D. W. Roberts, Rev. Sci. Instrum. **61**, 2914 (1990).

14. D. Wróblewski, K. H. Burrell, L. L. Lao, P. Politzer, and W. P. West, Rev. Sci. Instrum. **61**, 3552 (1990).
15. R. D. Cowan, The Theory of Atomic Structure and Spectra, University of California, Berkeley, 1981.
16. V. L. Jacobs and J. F. Seely, Phys. Rev. **A36**, 3267 (1987).
17. D. Wróblewski, L. K. Huang, and H. W. Moos, Rev. Sci. Instrum. **59**, 2341 (1988).
18. L. K. Huang, et al. Phys. Fluids B **2**, 809 (1990).
19. D. Wróblewski, K. H. Burrell, and R. Seraydarian, Bull. Am. Phys. Soc. **34**, 2116 (1989).
20. R. P. Seraydarian, K. H. Burrell, and R. J. Groebner, Rev. Sci. Instrum. **59**, 1530 (1988).
21. A. Boileau, M. von Hellermann, W. Mandl, H. P. Summers, H. Weisen, and A. Zinoviev, J. Phys. B: At. Mol. Opt. Phys. **22**, L145 (1989).
22. J. C. Kemp, G. D. Henson, C. T. Steiner, and E. R. Powell, Nature **326**, 270 (1987).
23. J. C. Kemp, J. O. S. A. **59**, 950 (1969).
24. L. L. Lao, et al., Nucl. Fusion **30**, 1035 (1990).
25. D. Wróblewski and L. L. Lao, Phys. Fluids B (in print).

X-RAY MEASUREMENTS FROM THE JET AND ASDEX TOKAMAKS

U. Schumacher[1], R. Barnsley[2], G. Fußmann[1],
K. Asmussen[1], C.C. Chu[1], G. Janeschitz[2]

[1] Max-Planck-Institut für Plasmaphysik, Association EURATOM-IPP,
8046 Garching, FRG

[2] JET Joint Undertaking, Abingdon, Oxfordshire OX14 3EA, England

ABSTRACT

The identification of impurities and the investigations into their behaviour in magnetically confined plasmas as well as the plasma parameter determination belong to the most important applications of x-ray spectroscopy in the JET and ASDEX tokamaks. Broad-band flat crystal monochromators - supported by impurity monitors and calibration devices - enable impurity concentration and plasma radiation loss measurements over a wide range of atomic numbers, while high-resolution spectroscopy offers plasma ion temperature and rotation measurements.

1. INTRODUCTION

The impurities in a magnetically confined plasma cause energy losses by line radiation mainly of medium and high Z elements and by bremsstrahlung. In addition the dilution of a D-T-plasma by light impurities is a major concern of future devices. Control of the impurity concentrations is hence an important task of present thermonuclear fusion research. Spectroscopy - especially in the soft x-ray region - offers a powerful tool for investigations into the impurity behaviour in the plasma, their composition and transport. This knowledge allows interpretation of the plasma radiation losses, the impurity distributions and the composition of the effective ion charge Z_{eff}. Moreover, spectroscopic information is normally also utilized for the determination of some of the most important plasma parameters such as temperatures and densities of ions and electrons or plasma motion[1-4]. Furthermore, detailed spectroscopic information gives possibilities to determine important atomic processes and to measure the related rate coefficients.

Likely impurities in a hydrogen, deuterium or helium plasma of a tokamak are the light elements Be, B and C from the near-plasma components or their layers, the medium-Z elements such as O, F, Ne, Al, Si, S, Cl and Ar, which are intrinsic impurities or which are deliberately injected into the plasma, as well as the metallic impurities such as Ti, Cr, Fe, Ni and Cu from the vacuum vessel walls or the divertor region.

Due to the impurity radiation losses strong efforts were made to reduce the metallic impurities in the plasma by covering the near-plasma components with low-Z material layers (carbonization, boronization and Be evaporation) or even replacing them by graphite or beryllium components, such that low-Z impurities are the dominant ones in present magnetically confined plasmas. These materials, however, show relatively high sputtering yields and low thresholds for sputtering which are especially disadvantageous for areas of high power fluxes such as those of the divertor plates. Hence high-Z materials like Mo or W with much higher sputtering thresholds are under discussion as materials for these components, such that the charge numbers Z of impurities possibly to be found in magnetically confined plasmas cover a wide range from the low to the very high.

These impurities emit K-, L- or M-shell radiation in the soft x-ray spectral region with wavelengths λ between about 0.1 nm and 10 nm or photon energies in the range of about 100 eV to 10 keV. In most cases, especially for the K-shell radiation, the spectral lines are strong and are found very well separated from each other or in groups. The hydrogen- and helium-like spectra are relatively simple and well understood. The wavelengths of the Lyman-α lines (1 s ^{2} S$_{1/2}$-2p ^{2}P°) of different elements of charge Z, for instance, scale approximately like $\lambda_{1s-2p} = 16\pi hc\varepsilon_o a_o/(3e^2 Z^2)$ - with the usual constants to give λ_{1s-1p} [nm] $\simeq$ 121.9/Z^2. From Z = 4 (Be with λ_{1s-2p} = 7.5928 nm) up to about Z = 36 (Kr) therefore these strong spectral lines can be found in the soft x-ray region, which allows an easy impurity identification.

The helium-like spectra are also relatively simple for any particular Z. They are emitted over a wide temperature range and hence play an important role in x-ray plasma spectroscopy[5,6].

The spectra of medium- and high-Z neon-like ions are useful for diagnostic purposes too, because they are prominent over a wide range of plasma parameters in tokamaks due to the closed shell configuration and the resultant stability of the ions. Therefore the radiation losses of these plasmas are dominated by the $\Delta n = 1$ transitions of the neon-like ions.

Moreover, neon-like ions are well suited to test calculations of the atomic structure of multi-electron systems. Because of the closed-shell configuration of their ground state their structure is simple enough for accurate calculations but simultaneously of sufficient complexity for theoretical model comparisons[7].

These applications of x-ray spectroscopy in magnetic confinement fusion call for quite different instrumentation. The impurity survey and the investigations of impurity behaviour need broad-band spectrometers or monochromators while most methods of plasma parameter determination from x-ray spectra and the investigations of important atomic processes rely on high resolution spectrometers.

2. BROAD-BAND X-RAY SPECTROSCOPY FOR IMPURITY INVESTIGATIONS

Broad-band soft x-ray spectroscopy is best performed with plane crystal monochromators, which fulfill the Bragg condition

$$n\lambda = 2d \sin\Theta \left(1 + \frac{4d^2 \delta}{n^2 \lambda^2}\right),$$

in combining the wavelength λ with the Bragg angle Θ. 2d is the crystal lattice constant, n the order of interference, and $\delta(\ll 1)$ is the refractive index minus unity. Depending on the accessible Bragg angle range a very small number of different crystals is sufficient to cover the above indicated wavelength (and photon energy) range completely for first order reflection (n = 1). In Fig. 1 the dotted lines indicate the Bragg angle limits of two flat crystal monochromators to be described below, and the spectral line groups to be observed ranging from hydrogen-like metals to helium-like beryllium are indicated at the top.

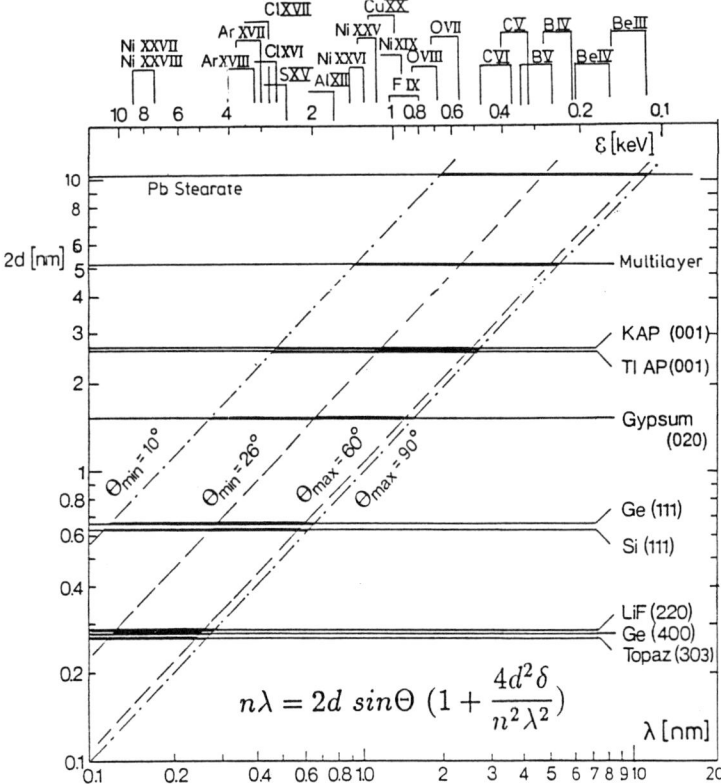

Fig. 1 The accessible spectral ranges in photon energy ε (upper scale) and wavelength λ (lower scale) for two typical flat crystal broad-band x-ray monochromators.

If the acceptance angle in the dispersive direction is limited by the crystal rocking curve width and the collimator acceptance angle to $\Delta\Theta$, the monochromator resolving power is given by $\lambda/\Delta\lambda = (1/\Delta\Theta)\tan\Theta$.

While most applications of crystal spectroscopy have been in the wavelength range below 2.5 nm the dominance of the low-Z impurities C, B or Be in present-day tokamaks calls for an extension into the longer wavelength region. Here, however, three major difficulties arise which all are related to the increasing absorption coefficients for these long wavelengths: The instrument sensitivity decreases due to absorption in the necessary thin foils, which have to separate different pressure regions such as that of the monochromator housing from the plasma vessel or from the detector pressure volume. The rocking curve widths of the crystals get very wide due to the linear absorption coefficient μ_L, such that the monochromator resolving power, which for a perfect lattice with absorption is given by

$$\lambda/\Delta\lambda \simeq \pi\lambda/(2\mu_L d^2),$$

decreases with a high power of the wavelength[8]. Moreover, the calibration of the instruments is more difficult for longer wavelengths than for shorter ones due to lack of calibration sources in that region.

2.1 SINGLE FLAT CRYSTAL MONOCHROMATORS

For a broad-band monochromator with a single flat crystal this crystal must be rotated at an angular speed of $d\Theta/dt$, while the detector has to be moved with twice this rotational speed. An example of this type is the "Multi-Bragg-Monochromator", which is sketched in Fig. 2.

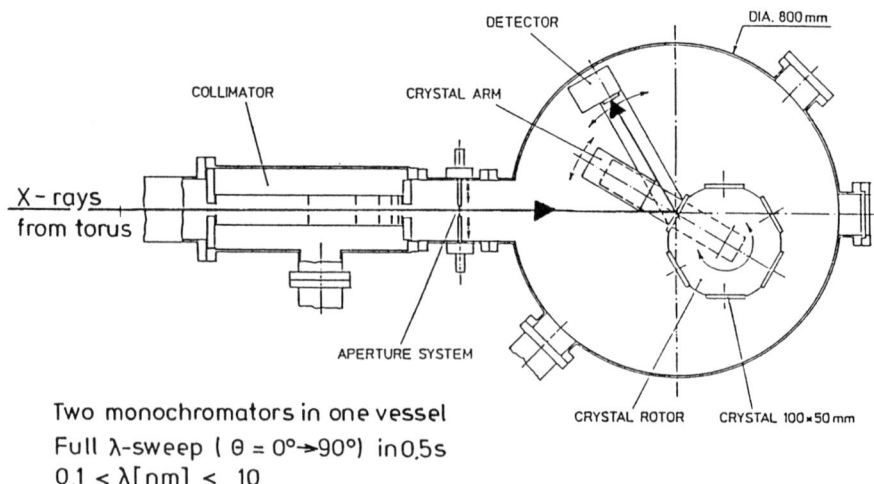

Fig. 2 The ASDEX Multi-Bragg monochromator

The crystal on its crystal arm scans the Bragg angle Θ at a rotational speed of up to $\pi\ s^{-1}$. The crystal arm carries the crystal rotor, which contains six different remotely changeable crystals, such that the soft x-ray wavelength range of interest - as indicated in Fig. 1 by the Bragg angle limits of 10° and 90° - can completely be covered. There are two monochromators in one vessel, which allow simultaneous wavelength sweeps with different crystals and different angular speeds and Bragg angle ranges to investigate different line groups at one time. A single instrument, however, can only integrate the radiation on one line of sight through the plasma. For ASDEX Upgrade therefore three Double-Bragg monochromators of this type are built which moreover can be moved up and down on a shot-to-shot basis to obtain spatial resolution of the radiation from the plasma.

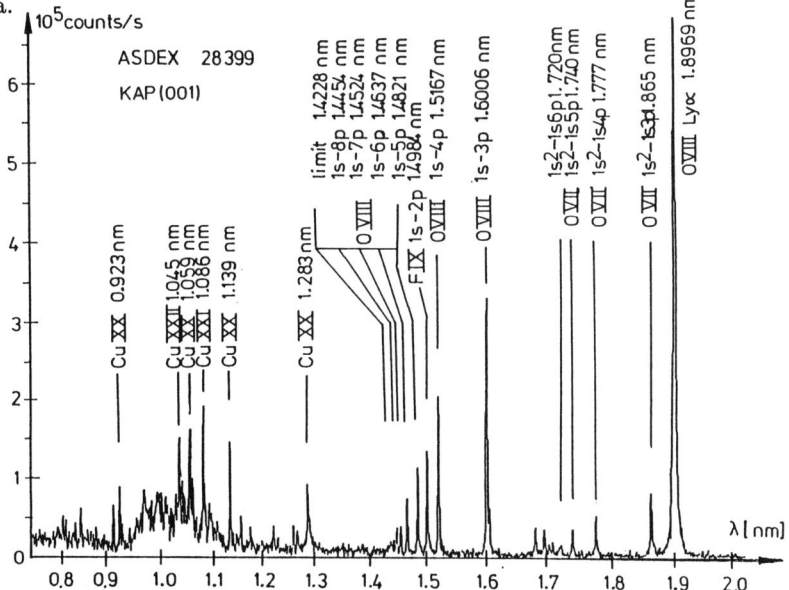

Fig. 3 Typical oxygen Lyman series broad-band spectrum of an ASDEX plasma obtained with the Multi-Bragg monochromator.

As an example of results obtained with this broad-band monochromator from ASDEX Figs. 3 and 4 give spectra of quite different discharges. In Fig. 3 the radiation is characterized by the Lyman series $1s\ ^2S_{1/2} - np\ ^2P°$ of oxygen OVIII, the corresponding series $1s^2\ ^1S_0 - 1snp\ ^1P_1$ of helium-like oxygen OVII and mostly neon-like copper lines. The ratios of the line intensities of the hydrogen-like oxygen lines mainly follow the values to be expected from purely collisional excitation by the plasma electrons as was discussed in ref. 11.

Fig. 4 gives the broad-band spectrum dominated by the L-shell transitions of metal impurities in an ASDEX discharge, following erosion of these metals from the divertor plates.

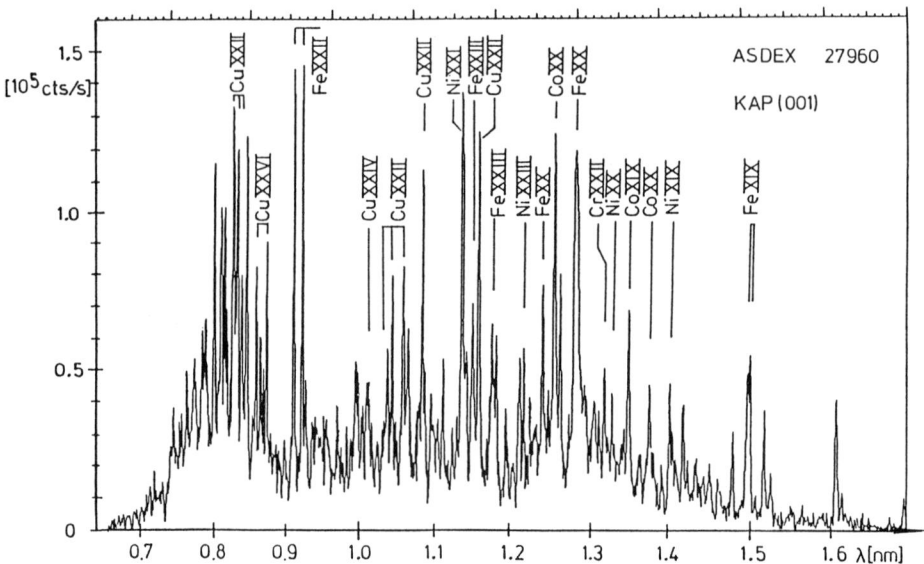

Fig. 4 Broad-band spectrum of an ASDEX plasma dominated by L-shell metal lines

The spectrum demonstrates the higher complexity of the - partly superimposed - L-shell spectra of the metal impurities as well as the importance of this wavelength region for the plasma radiation losses.

Broad-band spectra can also be obtained from flat-crystal rotors[9,10], which use a rotating hexagonal crystal holder and a large multiwire proportional counter at fixed location.

2.2 DOUBLE-CRYSTAL BROAD-BAND SPECTROSCOPY

Broad-band soft x-ray spectroscopy for the impurity survey of DT plasmas - as anticipated in the JET active phase - requires very good shielding of the detector against the high fluxes of neutrons and hard x-rays. Due to its fixed detector position which allows a labyrinth radiation shield a double-crystal monochromator is an appropriate instrument for active phase x-ray spectroscopy. The double-crystal monochromator at JET can perform wide spectral scans to cover - if the appropriate crystals are selected - the spectral range from about 0.1 to 2.3 nm, as can be deduced from Fig. 1 with the Bragg angle limits of 26° and 60° of this device[11]. If the important condition of parallelity of both crystals is met this relatively simple double-crystal device allows absolute calibration for sensitivity and wavelength. The instrument has demonstrated its routine operation to obtain broad band spectra for the impurity survey, to perform repetitive scans of spectral line groups to study line ratios or measure the Lyman decrement of hydrogen-like ions, to deliver line scans for line profile measurements with moderate time resolution to determine ion temperatures and plasma rotation velocities. The instrument has been operated in the monochromatic mode for the determination of the time behaviour of the intensity of a specific spectral line for impurity transport investigations in relation to an impurity injection by gas puff, by laser ablation or by sinusoidally modulated gas inlet for harmonic analysis[11,16]. An example of a broad band spectrum obtained with this double-crystal monochromator is given in Fig. 5, where hydrogen- and helium-like chlorine spectral lines are dominating. Like most of the oxygen Lyman series spectra obtained with JET, the ratios of these chlorine lines also seem to follow the Lyman decrement characterized by dominating electron impact excitation of the upper levels of these lines. In contrast to spectra of the carbon Lyman series observed under conditions of strong contribution of charge-exchange recombination processes[12], these oxygen (and chlorine) spectra were in agreement with pure electron impact excitation[11].

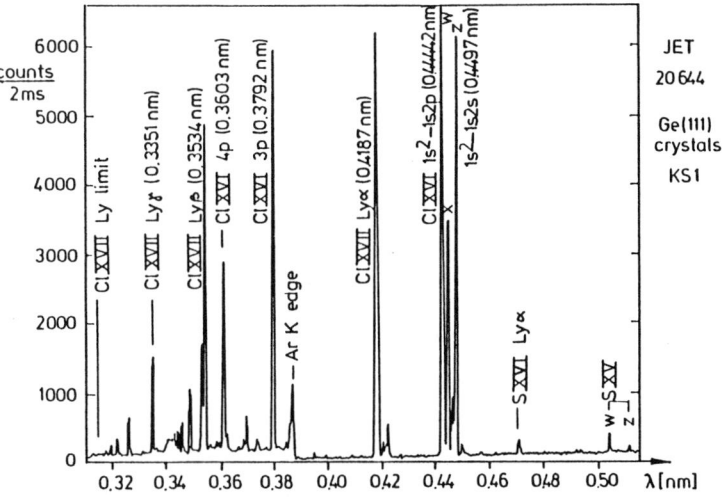

Fig. 5 Hydrogen- and helium-like chlorine line dominated broad band spectrum obtained with Ge(111) crystals in the double-crystal monochromator at JET. The ArK edge is observed because Ar is applied as detector gas.

Several members of the Lyman series can be observed; the maximum quantum number n^* could be limited either by Doppler broadening, Stark broadening or finite instrumental resolution. If the monochromator apparatus profile did not contribute to the observed line widths and the line separation were set equal to the full Doppler plus Stark width[13], the maximum observable quantum number n^* would be given by

$$n^* = (1.215 \cdot 10^{-3} \cdot (T_i \,[keV]/A)^{1/2} + \Delta\lambda_{Stark}(n_e)/\lambda)^{-1/3}$$

as indicated in the left part of Fig. 6, where A is the atomic weight, showing that the Stark widths $\Delta\lambda_{Stark}$ do not play a role for the electron density range below a few times 10^{20} m^{-3}, so we have about $n^* = 9.37/(T_i \, [keV]/A)^{1/6}$. However, the finite instrumental resolution is important. For the coarse collimator in the double-crystal monochromator, and applying TlAP and Ge(111) crystals, $n^* \simeq (\Delta\lambda_{App}/2\lambda)^{-1/3}$ and would be about 9 for oxygen and chlorine, respectively, as given in the right part of Fig. 6, in good agreement with the observations[11].

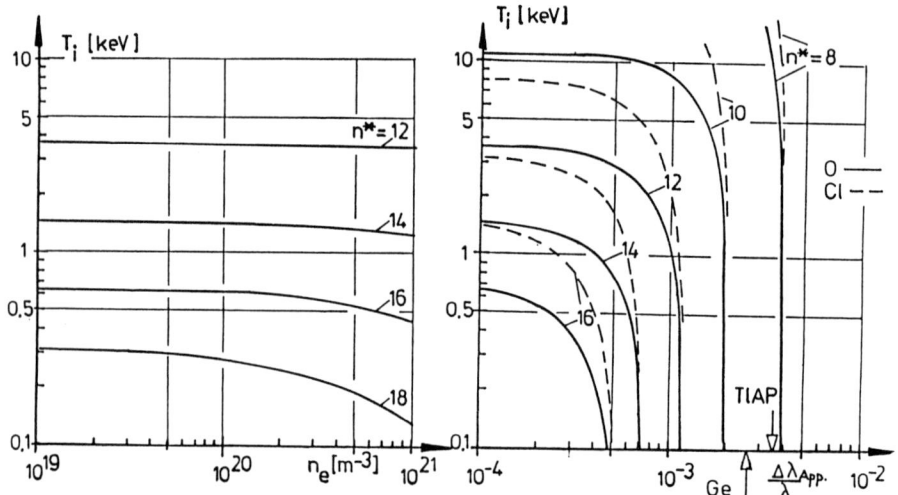

Fig. 6 The maximum observable quantum number n^* of hydrogen-like oxygen and chlorine as function of ion temperature T_i, electron density n_e and apparatus profile $\Delta\lambda_{App}/\lambda$.

The double-crystal monochromator gives absolute line intensities, from which impurity concentrations can be inferred, ion temperatures, rotation velocities and impurity transport characteristics. This latter information, however, needs spatial resolution of the spectra. This can be obtained from a double-crystal monochromator in which the first crystal can be rotated additionally about the optical axis.

2.3 CONTINUOUSLY SPACE-RESOLVING DOUBLE-CRYSTAL X-RAY SPECTROSCOPY

Rotating the first (plane) crystal of a double-crystal device about the optical axis between both crystals enables radiation to be accepted continuously in space. The Bragg condition is fulfilled for all these directions which lie on a cone with a full opening angle of $\pi - 2\Theta$, where Θ is again the Bragg angle, as sketched in Fig. 7.

If a part of this cone is brought into the poloidal plane of the torus the radial spectral line intensity distributions can be determined, as described in detail for the continuously space-resolved x-ray monochromator of JET[14]. However, the

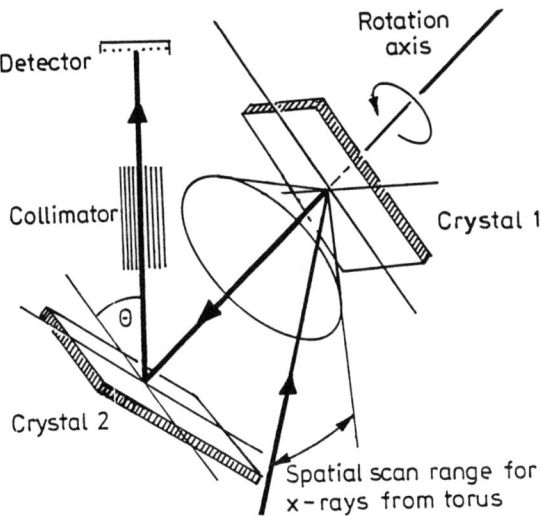

Fig. 7 Rotation of one of the crystals in a double-crystal device allows continuous spatial scanning.

swivel angle for this spatial scan should be limited because for large scan angles the spectral resolution and the instrument sensitivity go down[14,15].

At JET the radial intensity profiles of H- and He-like lines of Ni, Cl and Ar were taken. These distributions can be compared with the distributions calculated on the basis of the measured electron density and temperature distributions of JET and a diffusion coefficient of about $D = 1 m^2/s$, as was shown earlier[14].

From the calibration of this spatially scanning double-crystal monochromator with the large-area x-ray source at Ag Lα ($\lambda = 0.4155$ nm) by applying a silver coated anode the chlorine concentration, for example, is deduced to be about $n_{cl}/n_e \simeq 2 \cdot 10^{-5}$. The repetitive scans throughout the JET discharges allow to determine the time behaviour throughout every shot as well as the changes from shot to shot within the discharge series.

2.4 SPECIFIC X-RAY IMPURITY MONITORING

Since the broad-band impurity survey monochromators are flexible, cover a wide range of impurities and offer different investigations of the impurity behaviour they cannot always monitor the concentrations of specific impurities for every discharge. Hence simple x-ray monitors based on the intensity measurement of the Lyman-α (1s - 2p) transitions are needed for the concentration measurements of the most important low-Z impurities. Since these impurities are carbon and oxygen in ASDEX Upgrade, their Lyman-α-transitions (CVI 1s - 2p at $\lambda = 3.3736$ nm and OVIII 1s - 2p at $\lambda = 1.8967$ nm) are monitored by proportional counters measuring the x-ray beams collimated by multi-grid collimators and reflected in first order off Pb-stearate (2d = 10.04 nm) and KAP(001) (2d = 2.6632 nm) crystals, respectively.

The gridded collimator - with an angular acceptance of $\Delta\Theta \simeq (1/500)$ rad - is necessary to separate the lines in question from other spectral lines, which has to be checked with the survey impurity spectrometer. As is necessary for the impurity survey instruments, the C- and O-monitor has to be calibrated for absolute impurity concentration measurements. From the line intensity $I = \int \varepsilon d\ell$ an emission coefficient $\bar{\varepsilon}$ averaged over the emission layer of the line in question is inferred with the known electron density and temperature profile and impurity transport. $\bar{\varepsilon}$ is relatively insensitive to the latter two quantities, but proportional to the electron density n_e as well as to the density n_i of the ground state of the H-like impurity ion in the emission layer, $\bar{\varepsilon} = n_e n_i \cdot C_{ij}$, where C_{ij} is the electron collisional excitation rate with j = 2 as main quantum number, and contributions from charge exchange processes are neglected.

First oxygen Lyman-α line intensity measurements were taken from ASDEX Upgrade. The maximum count rate of $2 \cdot 10^5$ counts per s observed in some of the discharges corresponds to a few % of oxygen impurity concentration.

3. ABSOLUTE X-RAY CALIBRATION

Absolute impurity concentrations can only be deduced from absolute x-ray line intensity measurements. The sensitivity of the monochromators and impurity monitors described above depends on the crystal reflectivity, its rocking curve width, the accepted solid angle, the different transmission values and the detector efficiency. Although all these properties can be measured separately, the overall accuracy is desirable. Since the plasma is an extended x-ray volume source, the calibration device must be a large-area x-ray source. For the calibration source of $10 \times 30 cm^2$ anode area as described before[17], many additional K-, L- and even M-lines were calibrated utilizing different anode materials. The absolute, integrated line intensities (radiances) for an electron current of $I_e = $ 1 mA and an electron energy of $E_o = $ 20 keV are plotted in Fig. 8a versus the photon energy E_x.

Except for the Kα-intensities of Ti and Si these intensities are in good agreement with the calculated values from the formula of Green[18] for the number N of characteristic x-rays per electron of energy E_o [keV], impinging on the target. It gives a line intensity (radiance) of

$$I_{line} [Wm^{-2} sr^{-1}] = (1/4\pi) \cdot K_o \cdot (E_o - E_x)^{1.63} \cdot j \, [A/m^2] \cdot E_x \, [eV] \cdot f,$$

where K_o is the quantum efficiency - as shown in Fig. 8b - E_x is the minimum exciting voltage for these characteristic x-rays in keV, j is the electron current density and f represents the correction of absorption by the target itself[18].

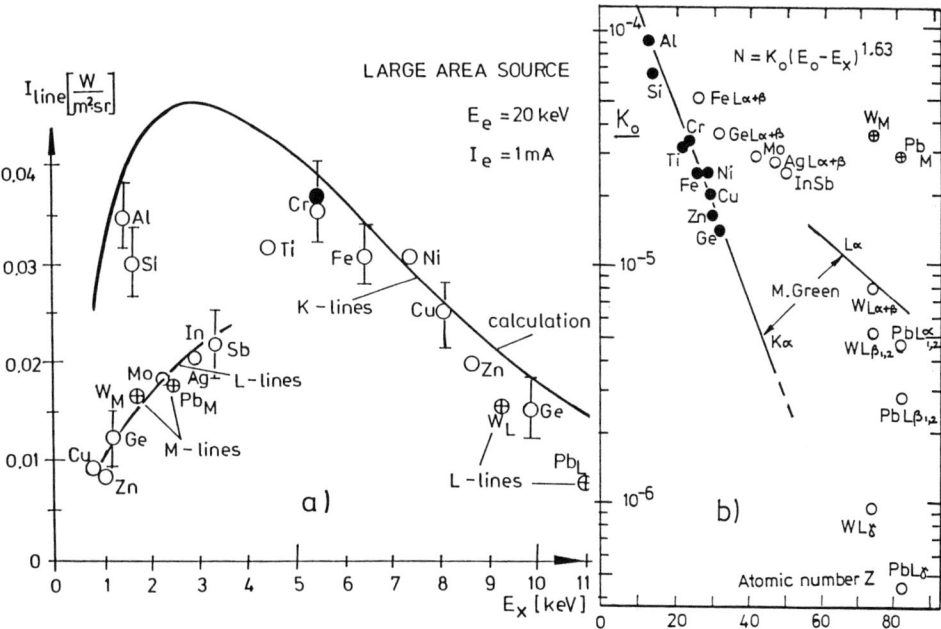

Fig. 8a) Absolute line intensities of K-, L- and M-transitions of different elements compared to the calculated values (solid line), b) the quantum efficiency values K_o as deduced from these intensities versus atomic number Z, added to and compared to values of M. Green[18] (solid lines).

The absolute calibration of this large-area x-ray source was obtained from the count rate of a Si(Li) detector for known solid angle and aperture area, supported by one measurement (solid dot in Fig. 8a) at MnKα utilizing a calibrated ^{55}Fe source. As a result an extension of quantum efficiency values K_o beyond those given by Green[18] is given in Fig. 8b. However, as mentioned before the absolute calibration for photon energies below about 1 keV is hampered by absorption processes in foils and detector.

4. HIGH RESOLUTION X-RAY SPECTROSCOPY

X-ray spectral line broadening and line shift measurements with high spectral resolution are widely applied to determine ion temperatures T_i and plasma rotation velocities from the Doppler width and the Doppler shift of x-ray lines, respectively[1,5,6,9]. At JET the ion temperature T_i is routinely deduced from the resonance line $1s^2\ ^1S_o - 1s2p\ ^1P_1$ of helium-like nickel (NiXXVII) utilizing a bent x-ray crystal spectrometer of the Johann type with about 10 m Rowland circle radius[19,20].

In order to determine the spatial distribution of the ion temperature T_i versus the minor radius r, in ASDEX the line of sight of a 1.5 m Rowland radius

Johann spectrometer was scanned throughout the poloidal plane[21]. Unfolding the measured distributions of the intensity

$$I(\lambda, r) = \frac{1}{2\pi} \int_r^a \varepsilon(r') \frac{e^{-[(\lambda-\lambda_o)/\Delta\lambda(r')]^2}}{\sqrt{\pi}\Delta\lambda(r')} \frac{dr'}{[1-(r/r')^2]^{1/2}}$$

gives the ion temperature distribution $T_i(r)$. λ is the wavelength with center at λ_o, $\varepsilon(r')$ the emission coefficient as function of minor radius r' and a the plasma outer radius. The results of the radial ion temperature profiles as obtained from the resonance lines $1s^2\ ^1S_o - 1s2p\ ^1P_1$ of helium-like Cl, Ar and Cu are plotted in Fig. 9. They are in relative good agreement, also in comparison to the results from the passive and the active charge exchange measurements[21]. The passive charge exchange diagnostic (circles) tends to overestimate the temperature for the outer plasma region.

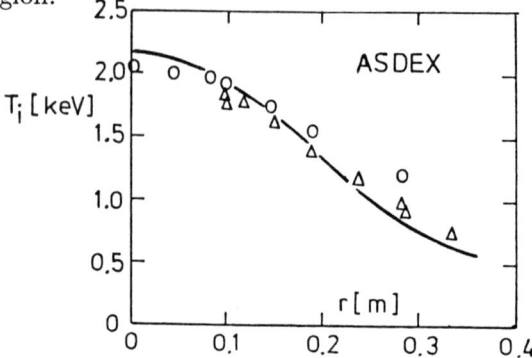

Fig. 9 The radial ion temperature distribution in ASDEX obtained from spatially resolved x-ray line profile measurements (solid line) in comparison to those from passive (circles) and active (triangle) charge exchange measurements

5. CONCLUSIONS

The x-ray measurements on the JET and ASDEX tokamaks have demonstrated that combining the tasks of extensive impurity investigations in magnetically confined plasmas with plasma parameter determination from x-ray line profiles can only be matched by utilizing flat crystal broad-band monochromators and impurity monitors as well as high spectral resolving curved crystal spectrometers.

6. ACKNOWLEDGMENTS

The authors are grateful to the JET and ASDEX teams for their support in operating the tokamaks. Special thanks go to Drs. R. Giannella, E. Källne, K. Krieger, F. Mompéan, H.W. Morsi, H. Röhr, G. Rupprecht and P. Thomas for help and many fruitful discussions and to K. Bethmann, H.E. Clarke, J. Fink, M. Hien, T. Patel, J. Ryan, G. Schmitt and B. Viaccoz for continuous support.

REFERENCES

1. S. von Goeler, M. Bitter, S. Cohen, D. Eames, K.W. Hill, D. Hills, R. Hulse, G. Lenner, D. Manos, P. Roney, N. Sauthoff, S. Sesnic, W. Stodiek, F. Tenney, J. Timberlake, Proc. Course on Diagnostics for Fusion Reactor Conditions, ed. P.E. Stott et al. (Comm. Europ. Commun., Brussels 1983) Vol. I, p. 109
2. C. de Michelis, M. Mattioli, Nucl. Fusion **21**, 677 (1981)
3. E. Källne, in Phys. of Highly Ion. Atoms, ed. R. Marrus (Plenum, 1989)
4. R. Barnsley, K.D. Evans, N.J. Peacock, N.C. Hawkes, Rev. Sci. Instrum. **57**, 2159 (1986)
5. R. Bartiromo, Proc. Course on Basic and Advanced Diagnostic Techniques for Fusion Plasmas, ed. P.E. Stott et al. (Comm. Europ. Commun., Brussels 1987), Vol. I, p. 227
6. M. Bitter, S. von Goeler, R. Horton, M. Goldman, K.W. Hill, N.R. Sauthoff, W. Stodiek, Phys. Rev. Lett. **42**, 304 (1979)
7. P. Beiersdorfer, Proc. 15th Int. Conf. on X-Ray and Inner-Shell Processes, UCRL-JC-103919, Lawrence Livermore Nat. Lab., 1990
8. A. Burek, Space Sci. Instrum. **2**, 53 (1976)
9. S. von Goeler, Diagnostics for Fusion Experiments, Ed. E. Sindoni and C. Wharton, (Pergamon, New York, 1979), p. 79
10. R. Barnsley, K.D. Evans, N.J. Peacock, N.C. Hawkes, Rev. Sci. Instrum. **57**, 2159 (1986); R. Barnsley et al., Ref. 1, p. 287
11. R. Barnsley, U. Schumacher, E. Källne, H.W. Morsi, G. Rupprecht, Rev. Sci. Instrum. **62**, 889 (1991)
12. M. Mattioli, N.J. Peacock, H.P. Summers, B. Denne and N.C. Hawkes, Phys. Rev. A**40**, 3886 (1989)
13. A. Poquerusse, Phys. Lett. 41A. 453 (1972)
14. U. Schumacher, E. Källne, H.W. Morsi, G. Rupprecht, Rev. Sci. Instrum. **60**, 562 (1989)
15. U. Schumacher. Nucl. Instrum. Methods A **259**, 538 (1987)
16. K. Krieger, G. Fußmann, ASDEX-Team, Nucl. Fusion **30**, 2392 (1990)
17. H.W. Morsi, H. Röhr, U. Schumacher. Z. Naturforsch. **42a**, 1051 (1987)
18. M. Green, X-Ray Optics and X-Ray Microanalysis, Academic Press. N.Y., London 1963, p. 185; M. Green, Proc. Phys. Soc. **83**, 435 (1964); M. Green, V.E. Cosslett, Brit. J. Appl. Phys. **1**, 425 (1968)
19. R. Bartiromo, F. Bombarda, R. Giannella, L. Panaccione, G. Pizzicaroli, Rev. Sci. Instr. **60**, 237 (1989)
20. K.-D. Zastrow, H.W. Morsi, M. Danielsson, M.G. von Hellermann, E. Källne, R. König, W. Mandl, H.P. Summers, accept. J. Appl. Phys.
21. C.C.Chu, R. Nolte, G. Fußmann, H.U. Fahrbach, W. Herrmann, E. Simmet, ASDEX-Team, Proc. Eur. Conf.on Contr. Fusion and Plasma Physics **15c**, IV, 297 (1991)

STUDIES OF HEATING AND IMPURITY TRANSPORT IN THE PLASMA BOUNDARY OF TOKAMAKS

G M McCracken and U Samm[+]

AEA Fusion, Culham Laboratory, Abingdon, Oxon. OX14 3DB, UK
(Euratom/UKAEA Fusion Association)

[+]Association KFA Euratom KFA, D-5170 Julich, Fed Rep Germany

1. INTRODUCTION

Impurities in fusion plasmas lead to a loss of reactivity by radiation and fuel dilution.[1] The impurity concentration in the plasma is determined both by the production rate and transport in the plasma, particularly the transport in the boundary layer.[1,2] In this layer impurities, usually starting as cold atoms or molecules, are ionised and heated while diffusing both parallel and perpendicularly to the confining magnetic field. Atomic physics plays an important role in these processes. In the present paper we consider some of the experimental results on impurity behaviour in this boundary and some simple models to describe their behaviour. The discussion will be on the behaviour of impurities in plasmas bounded by limiters.

2. FUELLING OF IMPURITIES; He, Ne and Ar

Impurities entering as neutral atoms have an ionisation mean free path determined by the local electron density $n_e(r)$ and temperature $T_e(r)$. For an atom with a velocity v_o in the radial direction, the local ionisation rate is:

$$S(r) = \frac{d}{dr}[F_o \, exp\{-\int n_e(r) \, \overline{\sigma v}(r)/v_o dr\}] \qquad (1)$$

where $\overline{\sigma v}$ is the ionisation rate coefficient for a maxwellian distribution, and F_o is the initial flux. From measured $n_e(r)$ and $T_e(r)$ profiles an estimate can be made of the impurity source function $S(r)$. In the case of molecules the situation is more complicated because molecular ionisation frequently occurs first, followed by parallel transport, dissociation and further ionisation.

Impurity injection experiments are a valuable method of studying impurity transport. Injection of rare gases into the TEXTOR tokamak has shown that the impurity concentration in the edge builds up linearly at short times and tends to a steady state value at times long compared with the effective particle replacement time τ_p^*.[3] The edge impurity concentration N can be described empirically by an expression applicable to a zero dimensional model

$$N(t) = S_o \tau_p^* + (N_o - S_o \tau_p^*) \, exp\,(-t/\tau_p^*) \qquad (2)$$

where N_o is the initial concentration and S_o is the total influx across the last closed flux surface (LCFS). The real particle replacement time is given by $\tau_p = \tau_p^*(1\text{-}R)$ where R is the recycling coefficient of the impurities at the limiter.

Experiments have been carried out introducing He, Ne and Ar to determine the effective fuelling rates. The time dependence of the plasma parameters during neon injection is shown in fig. 1. The rate of gas injection is constant for a period of 0.5 sec during the plasma pulse. The electron density is also maintained constant by a feedback system controlling the main gas species. The radiation from the neon, NeIII, increases almost linearly during the injection period and the total radiation and the central effective charge, Z_{eff}, increase correspondingly. When the impurity gas inflow is switched off the signals characteristic of the impurity decay exponentially with time during the period when the background density and plasma current remain constant. This behaviour is predicted by the simple 0-D model described above. There is the possibility that the time constants are largely determined by impurity transport in the plasma rather than by τ_p^*. However, independent experiments [4], using very short (20ms) He injection bursts, have shown that the measured time constant for transport to the centre was $\leq$ 100ms. Similar results have been obtained in TFTR.[5] The time constants observed in the present experiments for He, Ne and Ar are much longer than those obtained for transport in the plasma and this indicates that they must be determined by τ_p^* which in turn implies high recycling coefficients.

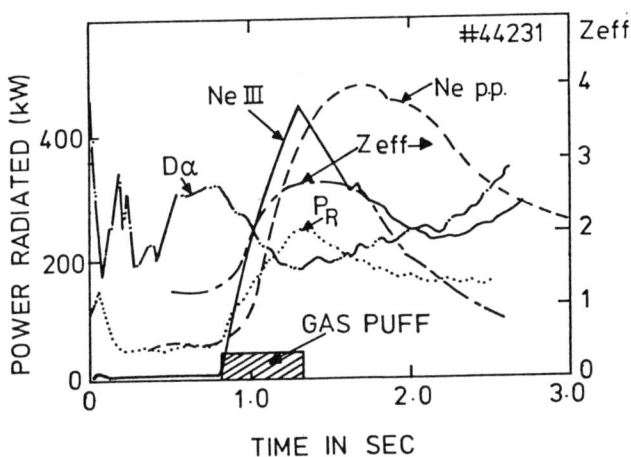

Figure 1: *Time dependence of $P_R, Z_{eff}, NeIII, D_\alpha$ and neon partial pressure during a 0.5s neon gas puff at the plasma wall in TEXTOR. $I_p^3 = 350kA$, $\bar{n}_e = 2 \times 10^{19} m^{-3}$.*

Measurements for a series of different gas puffing rates have been carried out for He, Ne and Ar. By measuring Z_{eff} and calculating the number of impurity atoms in the plasma (assuming Z_{eff} is constant across the plasma), we can obtain the ratio of the number of atoms in the plasma to the total number injected. This is a measure of the fuelling rate. Results from such experiments in TEXTOR show that while helium has a high fuelling rate, $\sim$ 100%, neon has a lower value, 40%, and argon is lower still, $\sim$ 3%.

The ionisation rate has been calculated as a function of radius for the three gases using equation 1 and the directly measured density and temperature profiles. The results are shown in fig. 2. It is clear that for a source at the

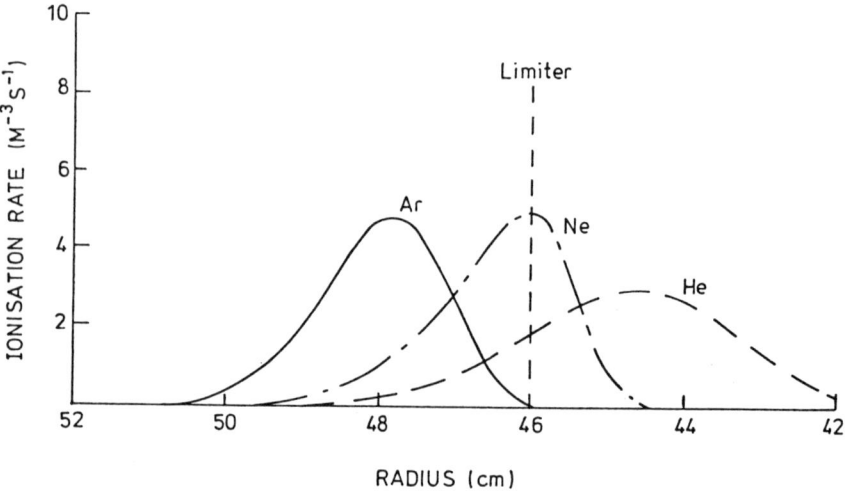

Figure 2: *Calculated ionisation rate of 0.05eV atoms starting from the wall (r = 52cm) using measured $n_e(r)$, $T_e(r)$ profiles for He, Ne and Ar.*

wall, the argon is ionised further out radially than the neon. The helium atoms are mostly ionised radially inside the last closed flux surface. The ionisation of the argon outside the LCFS is not, however, a complete explanation of the low impurity content in the plasma, since argon ionised outside the LCFS will in general hit the surface of the wall or limiter and recycle. By a series of such collisions it could eventually "walk" in to the LCFS. However, experiments comparing operation with the ALT II pumped limiter [6] on and off, has given direct evidence that argon is trapped in these surfaces more than neon or helium, fig. 3. Whereas when the pumped limiter is off there is no significant decay in the neon signals, it appears to make no difference to the pumping rate for argon whether the pumps are off or on. This shows that there is significant pumping of argon by the surface of wall or limiter. This is an important factor causing the low fuelling rate.

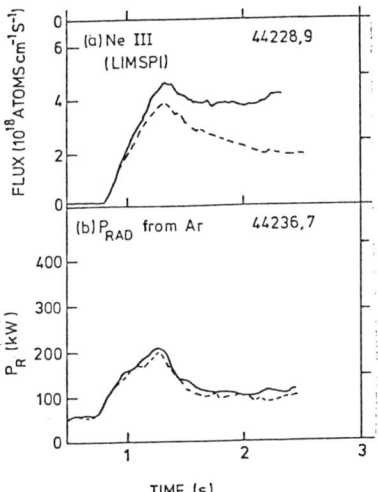

Figure 3: *Comparison of behaviour of (a) neon and (b) argon behaviour during and after gas puffing with ALTII open - - - - - and closed ———*

3. RESULTS FOR CO and CH$_4$

Global measurements show that for CO and CH$_4$ injection, the Z_{eff} increase is significantly lower than for neon with the same gas puff rate. Results for the time dependence of the main plasma parameters for CO injection are shown in fig. 4. The change in the carbon recycling rate at the limiter is not detectable above the noise level. A small rise in the oxygen level is detected. The rise in the radiation and Z_{eff} are much lower than in the case of neon. Although they are expected to be lower due to the slightly lower value of atomic number, it is found quantitatively that the fuelling rate of CO is <10%. The clear implication is that the recycling coefficient is much lower for carbon than for neon.

Direct observations of the point of injection show the parallel motion of the ionised species along the magnetic field. Some typical results are shown in fig. 5.[7] The spread of the neutral carbon CI is determined for given plasma conditions by the energy and angular distribution of the neutrals. The widths of the CII and CIII species are determined by parallel transport along the magnetic field direction and by the time required to ionise each species to the next higher state. The width decreases as the background plasma temperature increases.[8] The situation is more complex for the case of the molecular species,

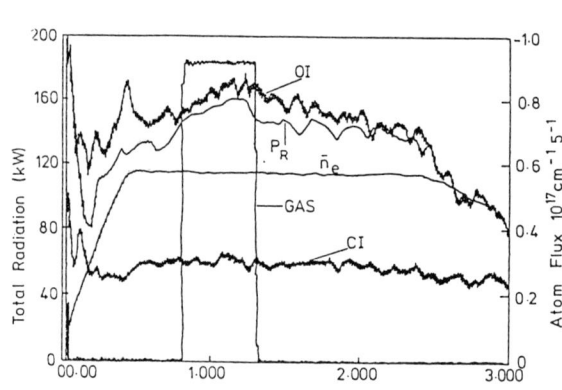

Figure 4: *Time dependence of radiated power, P_R, OI and CI at the limiter and electron density n_e during a 0.5 sec CO puff into a deuterium plasma in TEXTOR. $I_p = 350kA$, $n_e = 2 \times 10^{19} m^{-3}$.*

CO, than for the rare gases, since ionisation of the molecule is the most probable reaction.[9] The principle reactions for which there are data are shown in fig.

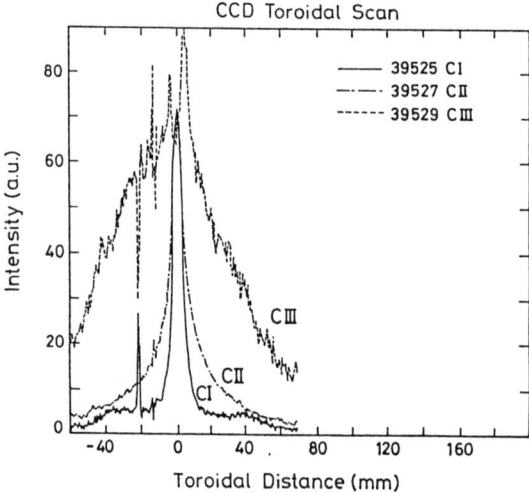

Figure 5: *Spatial distribution of carbon emission, CI, CII, CIII, along the magnetic field direction during CO puffing from a rail limiter at radius 46 cm in the TEXTOR tokamak.*

6 and compared with the cross–sections for ionisation of the atomic species.[10] Dissociative ionisation (to $C^+ + O$ or $O^+ + C$) has a cross-section which is a factor of 6 lower than simple ionisation for the electron temperatures of

interest.[9] After ionisation to CO^+, the molecular ion must move along the magnetic field until it is dissociated. The dissociation occurs rapidly ($< 1\mu s$) as the cross-section is larger than for the original ionisation of the molecule.[10] No direct information is available on the energy of resulting atoms or ions from the dissociation of the CO^+, although there is a suggestion that it will result from the same state as the CO neutral molecule.[11] Energy distributions from the neutral molecule have been directly measured and are typically $\simeq$ 0.05eV.[12]

The spreading in the toroidal direction has been modelled using the

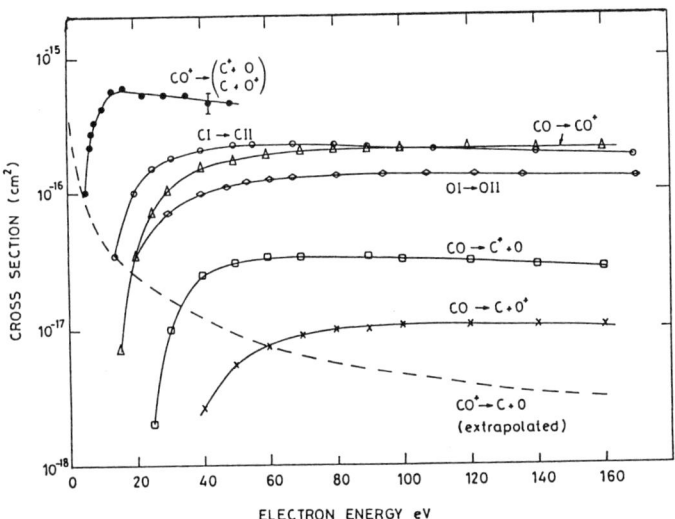

Figure 6: *Ionisation rates for various impact ionisation reactions in CO.*[9,10,11]

LIM Monte Carlo code.[13] The background density and temperature are taken directly from experimental measurements. Good fits to the experimental data can only be obtained if energies less than 0.1eV are used, consistent with the measured data for the neutral molecule.[12] The toroidal distribution of carbon from CH_4 and the oxygen and carbon from CO are compared in fig. 7. The spatial distributions are indistinguishable. This shows that the oxygen and carbon energies are approximately the same, as would be expected from momentum conservation and the nearly equal masses. It is also strong evidence that C from CH_4 has a low energy. This appears to be in contradiction with theoretical estimates which suggest that much higher energies may occur.[14] However, the complete dissociation of CH_4 is very complex, with many reaction chains, and the most probable energy is difficult to assess.

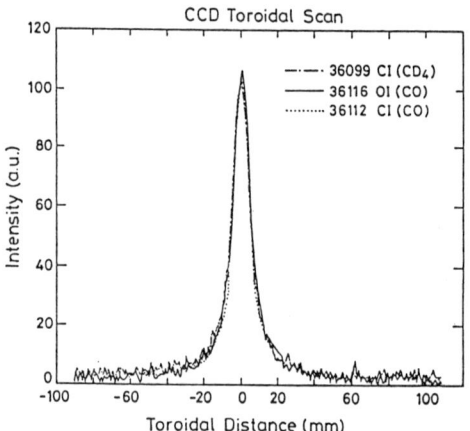

Figure 7: *The spatial distribution of CI from methane and CI and OI from CO during gas puffing through a limiter in the TEXTOR tokamak*

4. HEATING OF IMPURITY IONS

Impurity ions entering the plasma as gaseous species do so with low energy as discussed above. As soon as they are ionised, however, heating by ion-ion collisions starts. The amount of time an ion has in a given charge state before it is ionised to the next highest charge state is given by the ionisation time

$$\tau_{iz} = \frac{1}{n_e \overline{\sigma v}} \qquad (3)$$

The characteristic heating time is given by classical rates[15]:

$$\tau_{th} = \frac{m_I T_B^{3/2}}{1.4 \times 10^{-13} m_B^{1/2} n_B Z_B^2 Z_I^2 \ell n \Lambda} \qquad (4)$$

where m_I and m_B, Z_I and Z_B are the masses and charge states of the impurity and background plasma ions, n_B and T_B are the background plasma density and temperature and $\ell n \Lambda$ is the Coulomb logarithm. The thermalisation rate is

$$\frac{dT}{dt} = \frac{T_B - T}{\tau_{th}} \qquad (5)$$

where T is the impurity temperature. Using equation (5) we can calculate the impurity ion temperature attained in a given charge state, before ionisation to the next higher state

$$T = T_B - (T_B - T_o) exp(-\tau_{iz}/\tau_{th}) \qquad (6)$$

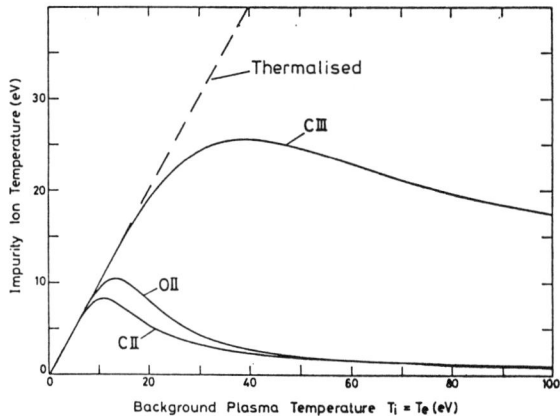

Figure 8: *Impurity ion temperatures, OII, CII and CIII calculated as a function of background plasma temperature in a deuterium plasma from equation 6.*

The ratio τ_{iz}/τ_{th} thus determines the fraction of the background plasma temperature which an ion gets to in a given charge state. It is seen that this is independent of density (assuming $n_B = n_e$) and dependent only on T_B (assuming $T_i = T_e$). The temperatures of a number of impurities, calculated on the basis of this simple model are shown in fig. 8. At low values of T_B, ionisation is slow but ion heating is rapid (being proportional to $T_B^{-3/2}$). Thus the impurity ions rapidly thermalize with the background plasma. On the other hand, at high plasma temperature ionisation is rapid and heating is slow. The impurity ions do not thermalize in the low ionisation stages.

Direct measurements of the impurity ion temperature during impurity gas puffing have been made by measuring doppler broadening with a high resolution spectrometer.[7] Although it is difficult to unfold the distribution, reasonable estimates of the temperatures can be made down to a few eV by folding together the instrument function, the structure of the Zeeman splitting in the known magnetic field, estimating an ion temperature and then fitting to the experimentally measured line width.

5. IMPURITY CHARGE STATE DISTRIBUTION

For impurities entering at a localized source it is clear that the low charge states will dominate close to the source. Ionisation to higher states takes longer and so the ions will have moved further from the source both toroidally, poloidally and radially.

With intrinsic impurities where the source is continuous throughout the discharge, the ions have time to diffuse into the centre and become highly

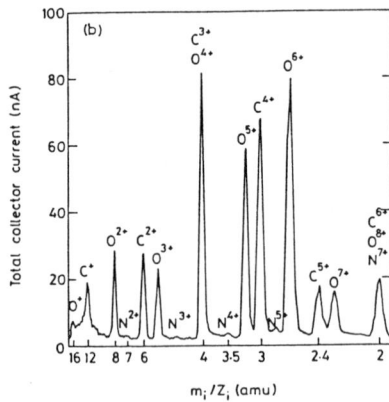

Figure 9: *Spectrum of mass/charge ratio for impurities in the DITE tokamak measured using a Plasma Ion Mass Spectrometer PIMS* [16] *in a deuterium plasma.*

stripped and heated. In equilibrium the inflow must be balanced by an equal flow outwards. Because recombination rates are slow compared with transport times, it is possible to get high charge states in the plasma boundary. This picture has recently been confirmed by measurements using a Plasma Ion Mass Spectrometer (PIMS).[16,17] A spectrum of mass/charge ratio is shown in fig. 9 for the intrinsic impurity ions in the DITE tokamak. Under the conditions studied, carbon and oxygen levels were high and approximately equal for the two species. It is observed that all charge states of the two impurities are present with the most probable ones being C^{4+} and O^{4+}. The charge state distribution has been compared with the LIM code calculations and rather good agreement obtained.[16]

6. SUMMARY AND CONCLUSIONS

Impurities entering a plasma are rapidly ionised and heated. These processes can be observed relatively easily with standard spectroscopic techniques. The observations also show transport along and across magnetic field lines. The heating and parallel transport are consistent with classical processes.

Ionisation in the plasma boundary can lead to collisions with the surfaces of wall and limiter. Where there is a high recycling coefficient (He, Ne) the impurities get into the plasma, probably by a random walk. This leads to long effective time constants determined mainly by the recycling coefficient R. The behaviour of such impurities can be described by a simple global model. Where the reflection coefficient is lower, e.g. argon and carbon, the effect is to reduce the fuelling rate in some cases by factors of ~ 100. This lower fuelling rate should affect the impurity concentration of intrinsic impurities, though this effect has not been widely recognised.

The situation with intrinsic impurities is more difficult to analyse because they have a spatially distributed source with a wide range of incoming ion energies and angles. However, these can also be analysed with Monte Carlo techniques. In the steady state intrinsic impurities can diffuse to the centre and become highly stripped. They can then diffuse out to the wall. Because recombination is generally slow compared with diffusion transport times, impurities with high ionisation states can be observed at the wall.

ACKNOWLEDGEMENTS

We are grateful to G Bertschinger, A Pospieszczyk, B Schweer and our other colleagues at KFA Julich and AEA Fusion, Culham Laboratory for their assistance in carrying out these experiments and their help in discussing these results. Thanks are also due to to P C Stangeby for many valuable discussions on the analysis and interpretation.

REFERENCES

[1] P C Stangeby and G M McCracken, Nucl Fusion 30, 1225 (1990).
[2] P C Stangeby, J. Nucl. Mat., 145-147, 105 (1988).
[3] G M McCracken et al, Proc.Eur.Phys.Soc., Berlin Vol IV, 245 (1991).
[4] D L Hillis, K H Finken et al, Phys. Rev. Lett. 65, 2382(1990)
[5] E J Synakowski, B C Stratton, P C Efthimion et al, Phys.Rev.Lett., 65, 2255 (1990).
[6] D Goebel, R W Conn et al., J.Nucl.Mat., 162-164, 115 (1989)
[7] G M McCracken et al., J.Nucl. Mat., 176 & 177, 191 (1990).
[8] C S Pitcher, P C Stangeby, D H J Goodall et al., J.Nucl.Mat., 162-164, 337 (1989).
[9] O J Orient and S K Srivastava, J Phys B, At Mol Phys 20 3923 (1987).
[10] K L Bell, H B Gilbody, J G Hughes et al, Culham Laboratory Report, CLM R216, (1982).
[11] J B A Mitchell and H Hus, J Phys B: At Mol Phys 18, 547 (1985).
[12] R Locht and J M Durer, Chem.Phys.Lett. 34, 508 (1975).
[13] P C Stangeby, C Farrell, S Hoskins and L Wood, Nucl. Fusion 28 1948 (1988).
[14] W D Langer and A B Ehrhardt, Fusion Technol. 15, 118 (1989).
[15] L Spitzer, The Physics of Fully Ionised Gases, Interscience (New York) (1956) 1956.
[16] G F Matthews et al, Plasma Phys. and Contr.Fusion, 31, (1989).
[17] G F Matthews, R A Pitts et al, Nucl. Fusion 31, 1495 (1991).

PLASMA PROCESSING

DIAGNOSTIC MEASUREMENTS IN RF PLASMAS FOR MATERIALS PROCESSING

J. R. Roberts, , J. K. Olthoff,
M. A. Sobolewski, R. J. Van Brunt, and J. R. Whetstone
National Institute of Standards and Technology
Gaithersburg, Maryland 20899

S. Djurović
Institute of Physics, Novi Sad,
Trg Dositeja Obradovića 4, 21000 Novi Sad, Yugoslavia

ABSTRACT

Radio frequency (rf) plasmas are utilized in many applications in materials processing, such as semiconductor fabrication, surface modification, and coating. Plasma processing has replaced older techniques, such as wet chemistry, because the latter could not provide the necessary characteristics as process demands increased. A good example of this is the reduction of the feature size in semiconductors. The present critical dimension for semiconductor processing is 0.8 μm and is anticipated to be $\leq$0.25 μm by the year 2000. At present only plasma processing exhibits the potential of producing these line widths.

An important factor, as the demands on the processing of materials become more critical, is exactly how to determine that the plasma is actually performing the process as designed. One way that is being investigated is to design control diagnostics that necessarily operate in real-time, *in situ*, without significantly perturbing the process. Many such diagnostic methods have been proposed and are vigorously being investigated. They include probing the plasma with lasers, electric and mass selecting probes, and by observing the emission of radiation coming from the plasma. All of these and others must be investigated if the demands of material processing are to be met. Some of the methods being investigated for process control diagnostics are presented.

I. INTRODUCTION

The use of various kinds of metrology to set the conditions of a complicated machine has been a goal of every manufacturer. For production line devices that utilize plasmas to process materials, such as semiconductor etchers, the norm has been to measure the applied voltage or power, the gas pressure, its constituency and flow rate. It has been observed[1] that this is not sufficient to guarantee the day-to-day reproducibility of the desired process or to match one "identical" process machine to another. Since the requirements on these process machines are becoming more and more demanding, e.g. with the decrease in feature size in semiconductors and the increase in throughput, it is necessary to find a set of process control measurements that will suit this new demand.

The purpose of the experiments described here is to investigate various measurements as diagnostics that are sensitive to plasma conditions, particularly changes in the electron and ion energy distributions that might find application in process controls. Since the free electrons in a plasma are primarily responsible for the population of excited states in atoms and ions, as well as the ionization of the atoms, one purpose of these experiments is to measure certain excited state populations (or population ratios) that are sensitive to changes in the electron energy distribution. Since the ion kinetic energy affects the reactions at the plasma/surface interface, a second purpose of this work is to measure correlations between changes in plasma parameters and the ion kinetic-energy distributions in the plasma.

The Gaseous Electronics Conference (GEC) RF Reference Cell[1] was chosen as the experimental platform because its plasma and electrical characteristics are being thoroughly studied at many laboratories throughout the U.S. Temporally and spatially resolved spectroscopy was chosen as the method to measure the optical emission from the plasma. This method is non-intrusive and has proven to be a successful diagnostic method for the investigation of plasma conditions. Energy analyzing mass spectrometry was chosen to measure ion energy distributions.

Mixtures of argon and helium were chosen to be the gases for the optical emission measurements because they will not react with each other or the plasma chamber producing unknown processes that could affect the interpretation of the experimental results. Also, since there exist a great deal of atomic data on these elements, such as emission wavelengths, transition probabilities and excitation cross sections, the experimental results are easier to interpret. The experimental conditions that were chosen were the same as the reference conditions of the Reference Cell (see Ref. 1). Previous work done by all the laboratories indicates that these were the conditions where the plasma was most stable and reproducible. The plasma parameters of the temperature and density for the electrons, ions, and atoms are expected to be in the range of similar plasmas.[2] For the mass spectrometry studies, argon was chosen because of the relatively simple chemistry involved in the discharge, the large amount of previous research done on argon plasmas, and the existence of a reasonably reliable set of cross section data.

Neutral helium triplet and singlet optical emission lines arising from the upper states 3p, 3d, 4s, 4p, and 4d were observed. The ratios of the triplet-to-singlet line intensities were recorded with changes in pressure and percentage of Ar in He. Radial scans of the plasma emission from argon lines were also observed. From these observations radial profiles of the relative populations of Ar excited states were deduced. Temporally resolved optical emission of Ar and Ar^+ were also observed as a function of pressure and distance from the powered electrode. The mass spectra reported here are for Ar^+ and Ar_2^+ utilizing the energy analyzing mass spectrometer. These observations were recorded as a function of probe position.

II. EXPERIMENTAL DESCRIPTION

A. Plasma Source

The experimental arrangement is schematically described in Figure 1. The GEC RF Reference Cell is a parallel plate discharge chamber with 102 mm electrodes separated by 25 mm. They are cylindrically symmetric and their surfaces are horizontal. The top electrode has 169 holes with a diameter of 380 μm to provide a showerhead gas inlet and is grounded to the chamber on the outside of the vacuum interface. The bottom electrode is powered by a 13.56 MHz rf power supply with a capacitively coupled matching network. Flow rates were 20 standard cubic centimeters per minute (sccm) and peak-to-peak voltages ranged from 40-200 V at a frequency of 13.56 MHz.

The cylindrical vacuum chamber is constructed of stainless steel and has 8 radially-looking side ports at the chamber midplane. Two of the ports are fitted with 136 mm diameter quartz windows for the spectroscopic observations. Two ports orthogonal to these are 152 mm conflat flanges to accommodate a turbomolecular pump and the radially translating probe tip of the mass spectrometer. Four 70 mm conflat flange ports at 45° with respect to the other 4 ports have mounted pressure transducers and an electric probe, which can be inserted into the plasma at its midplane.

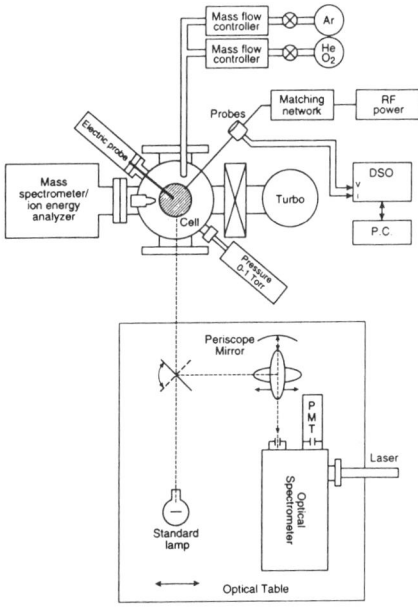

Figure 1. Schematic of experimental setup showing the Reference Cell, the energy analyzing mass spectrometer, electric probe, pressure gauge, gas flow control, power supply and matching network. Also shown is the optical spectrometer, reflecting periscope imaging optics, standard lamp, and components which are moved to scan emission profiles.

B. Optical Emission Measurements

The spectroscopic apparatus consists of a 2/3 m Czerny-Turner type grating spectrometer. This spectrometer is equipped with a cooled Burle* C31034A photomultiplier for detection of the emission signal. Both pulse counting and current mode observations were made. The pulse counting technique, using a time-to-amplitude-converter (TAC) and a multichannel-analyzer (MCA), was used for the temporally resolved measurements and is schematically shown in Figure 2. For current mode operation a picoameter is substituted for the pulse counting acquisition system. The spectrometer is equipped with a retractable mirror near its exit slit so a He-Ne laser may be substituted for the detector for alignment purposes. The vertical spectrometer entrance and exit slits are typically 150 μm wide and 2 mm high. The optics to image the plasma onto the spectrometer slit are front surface mirrors with coatings to efficiently reflect the plasma emission at wavelengths from 200 nm to 1500 nm. There are three flat mirrors and one concave mirror, each 152 mm in diameter. The concave mirror (650 mm focal length) is positioned so the plasma image is demagnified onto the entrance slit by approximately a factor of 2. These mirrors are arranged to act as a periscope bringing the level of the plasma emission to the same height as the spectrometer, as well as rotating the image of the plasma by 90°. By this rotation, the electrode surfaces are imaged parallel to the long dimension of the entrance slit, thus permitting observations close to the electrode surface as well as providing the highest possible spatial resolution of the plasma to be observed. The spatial resolution was ~0.5 mm vertically and ~4 mm horizontally. Because of the periscope, scanning of the plasma emission between the electrodes can be accomplished by translating one of the mirrors (see Figure 1). Translating the table permits horizontal scans of the plasma and an Abel inversion calculation[3,4] is performed to convert the horizontal emission profile into a radially symmetric distribution of the plasma emission. This process is necessary if real spatial distributions are to be interpreted. A calibrated tungsten filament lamp is mounted on the optics table and is substituted for the plasma source by rotating one of the flat mirrors (see Figure 1). This lamp is used to calibrate the optics/spectrometer system to obtain absolute spectral radiance measurements.

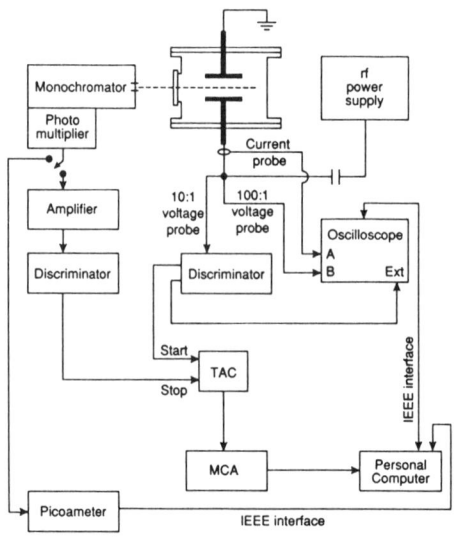

Figure 2. Schematic diagram showing optical emission data acquisition systems. The temporally resolved measurements are accomplished with a pulse counting technique utilizing the TAC-MCA and digital oscilloscope recording via an IEEE interface to a personal computer. The time-integrated measurements are done substituting a picoameter.

C. Energy Analyzed Mass Spectrometer

The mass spectrometer apparatus is a VG Instruments*, SXP300-CMA system which consists of a cylindrical mirror ion-energy analyzer[5] coupled to a 300 atomic mass unit (AMU) quadrupole mass spectrometer. Ions are sampled via a 200 μm aperture through a grounded stainless steel cone into the differentially pumped region of the analyzer. The orientation of the analyzer and sampling orifice with respect to the cell electrodes is shown in Figure 3.

Note that ions are sampled from the side of the plasma and not through an electrode as is common in other experiments.[6-8] This configuration is thought to be closer to the geometry which would be employed if this diagnostic were installed in a commercial etching reactor. The ion-energy analyzer and mass spectrometer may be moved by means of a bellows assembly so the distance from the sampling orifice to the edge of the electrode assembly may be varied from 0-10 cm. Ion kinetic energy distributions are obtained by selecting a particular mass from the spectrometer and then scanning the energy of the ions entering the energy analyzer. An energy resolution of 0.5 eV is maintained over the entire energy range scanned. Adjustments to the ion energy scale to correct for surface charging effects in the sampling cone are made based upon the observed kinetic-energy threshold for Ar^+ detection from an argon discharge.

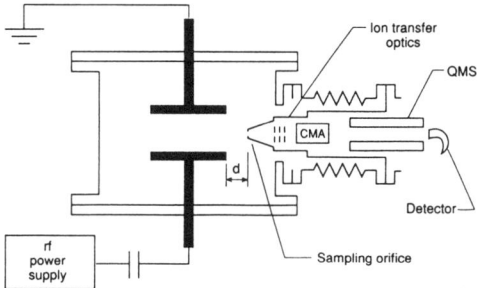

Figure 3. Schematic diagram showing the orientation of the ion-energy analyzer and mass spectrometer with respect to the Reference Cell electrodes. The distance from the edge of the electrode assembly to the aperture is d, the sampling orifice is a 200 μm, CMA is the cylindrical mirror ion-energy analyzer, QMS is the quadrapole mass spectrometer.

III. EXPERIMENTAL OBSERVATIONS

A. Optical Emission Measurements of He Triplet/Singlet Ratios

We observed emission from both the triplet and singlet lines of neutral helium. Triplet-to-singlet emission line ratios can be used to investigate the effect of the electron energy distribution on the excitation function.[9] The levels and the observed emission wavelengths in this experiment are shown in Figure 4. Transitions to the 1s ^1S ground state were not observed. The metastable 2s ^3S state of the He atom acts as the ground state for the triplet system. The population of excited states making transitions to the ground state and to this metastable level may exhibit radiation trapping effects. Therefore, a model of the excited state population is necessary to interpret emission intensities.[9] A detailed analysis may not be so important, however, if one only desires information concerning the temporally and spatially dependent electron energy distribution in the plasma, since some emission lines are more sensitive to changes in the electron energy distribution than others.

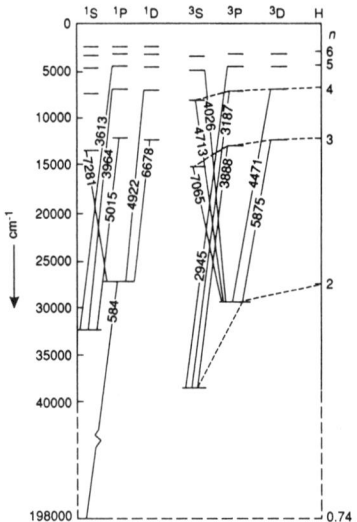

Figure 4. Grotrian diagram of He, with terms of H for comparison. Wavelengths are in Å. (H. G. Kuhn, <u>Atomic Spectra</u>, Academic Press, New York, 1962, p. 131).

Figures 5 and 6 show examples of the ratios of the triplet-to-singlet lines observed in this experiment as a function of pressure and percent of Ar admixed in He. Table I gives the correspondence between the upper state 3s, 3p, 4s, 4p and 4d and the line wavelengths in nm and their respective triplet and singlet designation (see Figure 4).

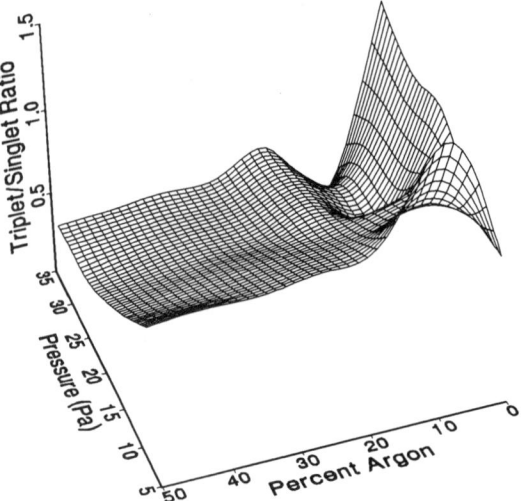

Figure 5. Optical emission measurements of the He 3p triplet-to-singlet ratio as a function of percent Ar in He and different pressures.

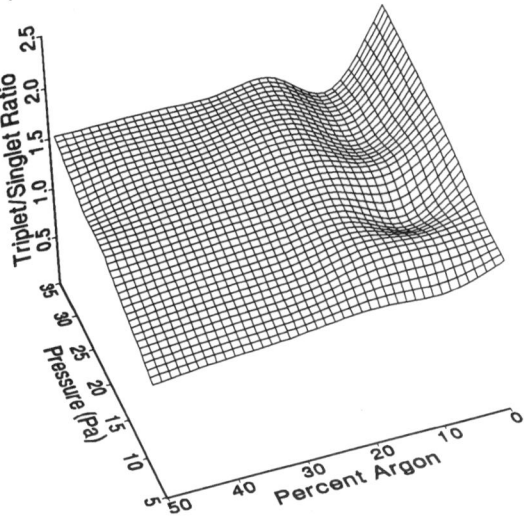

Figure 6. Optical emission measurements of the He 4d triplet-to-singlet ratio as a function of percent Ar in He and different pressures.

Table I Wavelengths (nm) and designations of measured He line ratios

3s	3p	4s	4p	4d
λ706.5(^{3}S)	λ388.9(^{3}P)	λ471.3(^{3}S)	λ318.7(^{3}P)	λ447.1(^{3}D)
λ728.1(^{1}S)	λ501.5(^{1}P)	λ504.7(^{1}S)	λ396.4(^{1}P)	λ492.1(^{1}D)

The observations of these He lines were taken over a range of pressures and admixtures of Ar. The pressures were: 6.67, 10.00, 13.33, 20.00, 26.66, and 33.33 Pascals (133.322 Pa/Torr). The percent of Ar in the total was: 0, 10, 25 and 50.

Since the ionization potential of Ar is much smaller than that of He, one expects small admixtures of Ar to perturb the conditions of the plasma. This can be seen by the gradients in the triplet-to-single ratio, especially for the 3p ratios (Figure 5). Because the population of the triplets and singlets are essentially two relatively independent systems in the He atom and the excitation rates have a different dependence on the electron energy distribution, these ratios may indicate changes in this distribution. The interpretation of this observation in terms of the change in the electron energy distribution function awaits a more thorough analysis of the population growth and decay processes.

B. Temporally Resolved Optical Emission Measurements of Ar and Ar$^+$

Temporally resolved observations of neutral Ar at 750.39 nm and Ar$^+$ at 434.81 nm were accomplished in a pulse counting mode using a TAC and an MCA shown in Figure 2. The TAC converts the time between the start of an rf cycle and the arrival of a photon pulse to a proportional voltage. This voltage is digitized by an MCA with

512 channels and a total acquisition time of 200 ns, thus giving a time resolution per channel of approximately 0.4 ns. The time history of both emission line signals were recorded with respect to the same time on the voltage waveform, therefore their relative phases may be compared.

In an optically thin medium the optical emission intensity per unit time in 4π steradians from a transition between two levels is given by,

$$I = n_u A_{u\ell} h \nu_{u\ell} \qquad (1)$$

where n_u is the upper state density, $A_{u\ell}$ is the atomic transition probability,[10] h is Plank's constant and $\nu_{u\ell}$ is the frequency of the transition. Figures 7 and 8 show, respectively, the temporal evolution of Ar and Ar^+ as a function of the distance above the powered electrode. The plot of the Ar^+ line shows a significant concentrations of ions near the powered electrode and as the distance between the electrodes is scanned the temporal evolution of the ion excited state density shows a significant distortion from sinusoidal. This is not observed in the Ar line. Here the neutral atom excited state density is very small near the powered electrode (the dark space) then rises abruptly after a few mm into a very bright region (the sheath) and decays more rapidly toward the grounded electrode. Also the neutral excited state density does not show any significant non-sinusoidal temporal evolution as does the ion excited state density. The Ar^+ signal has a much smaller signal-to-noise ratio than the neutral Ar which accounts for some of the bumpy structure in Figure 8. The interpretation of these temporally resolved optical emission signals requires the inclusion of many atomic processes. If one assumes the simple collisional-radiative model[9,11] of the plasma processes, then the population growth and decay processes may be assessed.

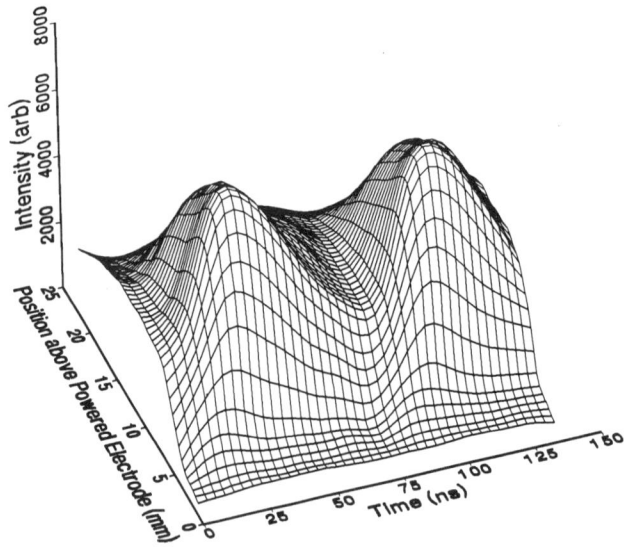

Figure 7. Optical emission measurements of the spatial profile and temporal evolution of the Ar 750.39 nm emission line. The distance between the electrodes is 25.4 mm and the period of the 13.56 MHz RF is 73.7 ns. The pressure of pure Ar is 13.33 Pa.

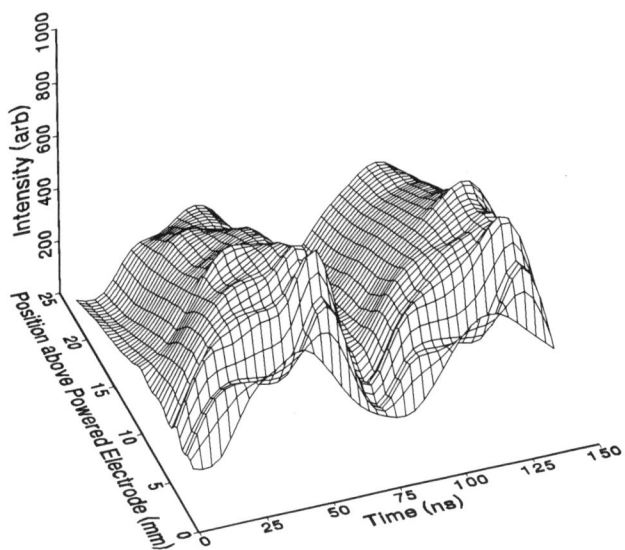

Figure 8. Optical emission measurements of the spatial profile and temporal evolution of the Ar⁺ 434.81 nm emission line. The distance between the electrodes is 25.4 mm and the period of the 13.56 MHz RF is 73.7 ns. The pressure of pure Ar is 13.33 Pa.

C. Spatially Resolved Optical Emission Measurements of Ar

One of the most desirable characteristics of a plasma processing device is to have a well known density profile of the plasma constituents, e.g. a constant radial ion distribution for plasma etching. To investigate the effect of changes in plasma parameters on the radial Ar emission profile, the horizontal distribution of neutral Ar line at 415.86 nm was observed in a pure Ar plasma as a function of pressure. This line was chosen because its emission intensity was much weaker than the stronger Ar emission lines; therefore radiation trapping was not a factor in the analysis. The plasma was scanned at the electrode midplane in a direction parallel to the electrode surfaces by moving the optical table supporting the spectroscopic apparatus horizontally. This scan was then inverted into a radial distribution of the optical emission signal by the Abel integral process.[3,4] The normalized signal of the data as well as the Abel inverted results are presented in Figure 9. The extent of the electrodes is also indicated for comparison. As can be seen the Abel inverted profiles show a substantial difference from the horizontally scanned data as the pressure increases. Also the plasma extends outside the electrode region. These results indicate at lower pressures the plasma density profiles can be expected to be the most uniform.

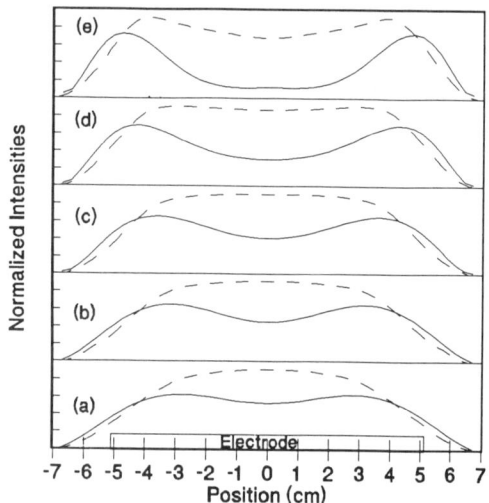

Figure 9. Abel inverted radial emission profiles of the Ar 415.86 nm emission line (solid) and the normalized horizontal emission intensity (dashed) for (a) 6.67, (b) 13.33, (c) 33.33, (d) 66.66, and (e) 133.32 Pa pressures. The electrode limits are indicated.

D. Energy Analyzed Mass Spectrometry

Shown in Figure 10 are the ion kinetic energy distributions for Ar^+ produced in an argon discharge as a function of the probe distance, d, from the edge of the electrodes. For small d the distributions exhibit a periodic structure similar to that observed previously for ion energy distributions sampled through the grounded electrode of parallel plate reactors.[8,12,13] This is expected since a sheath develops around the grounded cone as it moves closer to the electrodes thus suggesting that the cone behaves as an extension to the grounded electrode. The observed structure has been attributed to modulation effects associated with Ar^+ formation by resonant charge transfer in the sheath.[12] As the distance between the probe aperture and the electrodes increases, the structure disappears due to the diminishing influence of the probe in defining a sheath region near the aperture.

Ions sampled in the present configuration obtain most of their kinetic energy from traversing the sheath potential as they are accelerated from the bulk plasma toward the grounded surface. The maximum observed kinetic energy of an ion therefore provides an indication of the sheath potential if the energy distribution has not been severely disturbed by collisions in the sheath or by ions being created in the sheath region. The broad range of energies (0-20 eV) observed for Ar^+ in Figure 10 indicate that such processes do occur for Ar^+. Similar ion kinetic energy distributions are observed for Ar^{++} since Ar^{++} is formed in the sheath by high energy electron collisions.

Unlike Ar^+, Ar_2^+ ion kinetic energy distributions (see Figure 11) are narrow and exhibit no secondary structure. This indicates Ar_2^+ ions are not created in the sheath and the maximum ion energy may be used to estimate the sheath potential.[6] The distributions in Figure 11 are in agreement with previous ion energy distributions for Ar_2^+ measured by Köhler et al.[6] using a spherical energy analyzer sampling through a grounded electrode. As d increases the average energy of the Ar_2^+ ions decreases due to the influence of inelastic collisions.

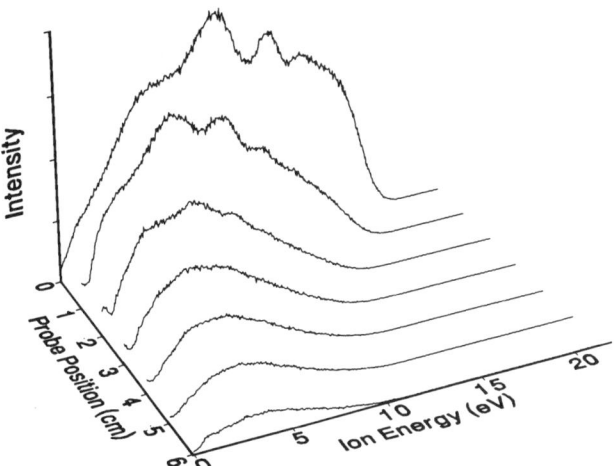

Figure 10. Ar^+ kinetic energy distributions as a function of probe position for a 13.33 Pa argon plasma with $V_{pp} = 200$ V.

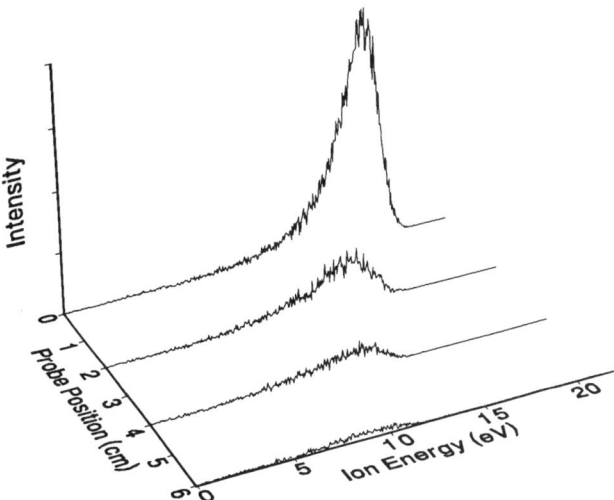

Figure 11. Ar_2^+ kinetic energy distributions as a function of probe position for a 13.33 Pa argon plasma with $V_{pp} = 200$ V.

IV. CONCLUSIONS

A substantial amount of information about an rf plasma may be derived using optical emission diagnostics and ion energy analyzed mass spectrometry. Optical emission spectroscopy provides a direct means of monitoring the spatial profiles of the density of plasma atoms and ions. Temporally resolved measurements allow one to deduce information about the evolution of these constituent densities present in the plasma. Investigation of the variations of different transitions allows one to monitor the effects of electrons as the plasma conditions are changed. Mass spectrometry with kinetic energy analysis can determine relative ion fluxes, monitor sheath potentials, and provide information about the interactions of ions and neutrals in the sheath region.

While a great amount of information is provided by these two methods, much of the data becomes extremely complex for plasmas on commercial etching reactors. Significant research must still be done in order to determine which aspects of the data are of direct importance in monitoring the plasma conditions and/or the etching process. An extension of the results presented here to systems containing etching gases and to more complex etching instrumentation is also required.

ACKNOWLEDGEMENTS

This work has been partially supported by SEMATECH.

REFERENCES

1. J. R. Roberts, J. K. Olthoff, R. J. Van Brunt, and J. R. Whetstone, <u>Advanced Techniques for Integrated Circuit Processing</u>, 1392, 428, (Society for Photo-Optical Instrumentation Engineers, 1991).
2. A. Picard, G. Turban and B. Grollean, J. Phys. D, 19, 991 (1986).
3. M. P. Freeman and S. Katz, J. Opt. Soc. Am. 50, 826 (1960) and J. Opt. Soc. Am. 53, 1172 (1963)
4. S. I. Herlitz, Ark. Fys. 23, 571 (1963).
5. H. Z. Sar-el, Rev. Sci. Inst., 38, 1210 (1967).
6. K. Köhler, J. W. Coburn, D. E. Horne, and E. Kay, J. Appl. Phys., 57, 59 (1985).
7. W. M. Greene, M. A. Hartney, W. G. Oldham, and D. W. Hess, J. Appl. Phys., 63, 1367 (1988).
8. M. F. Toups and D. W. Ernie, J. Appl. Phys. 68, 6125 (1990).
9. R. W. P. McWhirter, <u>Plasma Diagnostic Techniques,</u> edited by R. H. Huddelstone and S. L. Leonard (Academic Press, New York, 1965), Chap. 5, pp 201-264.
10. W. L. Wiese, M. W. Smith, and B. M. Miles, <u>Atomic Transition Probabilities,</u> Natl. Stand. Ref. Data Ser., Natl. Bur. Stand. (US) 22 Vol.II (1969).
11. C. Böhm and J. Perrin, J. Phys. D. 24, 865 (1991).
12. C. Wild and P. Koidl, J. Appl. Phys., 69, 2909 (1991).
13. J. Liu, G. L. Huppert, and H. H. Sawin, J. Appl. Phys., 68, 3916 (1990).
14. L. J. Overzet, J. H. Beberman, and J. T. Verdegen, J. Appl. Phys., 66, 1622 (1989).
15. C. W. Jurgensen, J. Appl. Phys., Vol. 64, 590 (1988).

* The identification of commercial materials and their sources is made to describe the experiment adequately. In no case does this identification imply recommendation by the National Institute of Standards and Technology, nor does it imply that the instrument is the best available.

ASTROPHYSICS

THE VARIABILITY OF ELEMENTAL ABUNDANCES IN THE UPPER SOLAR ATMOSPHERE

By
Uri Feldman

E. O. Hulburt Center For Space Research
Naval Research Laboratory Washington DC 20375-5000

Abstract

Coronal elemental abundances are found to change by as much as an order of magnitude relative to those present in the solar photosphere. Observations of modifications in coronal elemental abundances are reviewed and a tentative model governing the changes is discussed.

I. INTRODUCTION

Measurements of Solar Wind (SW) and Solar Energetic Particles (SEP) provided significant insight into the composition of solar elemental abundances. It was discovered (i.e. Breneman and Stone 1985, Gloeckler and Geiss 1989, and Coplan et al. 1990) that photospheric abundance ratios, between elements with low First Ionization Potential (FIP) and elements with high FIP, differs by a factor of 3-4 from ratios obtained in SW, and SEP (see Fig. 1). Postulating that the SW and SEP reflect the coronal composition, it was deduced that coronal and photospheric elemental abundances are different. Veck and Parkinson (1981), and Meyer (1985) reached a similar conclusion from the analysis of coronal spectroscopic data. Guided by these results, existing high quality data bases were analyzed in an attempt to reevaluate the elemental abundances in diverse parts of the solar upper atmosphere.

Some of the latest findings in the variability of the upper solar atmosphere are reviewed below.

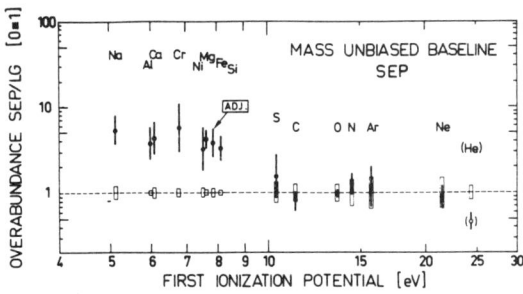

Fig. 1 SEP derived abundances relative to photospheric abundances, adopted from Meyer (1985).

© 1992 American Institute of Physics

II. SPECTROSCOPIC TECHNIQUES

The sun is composed of ≈90% H, ≈10% He, and traces of heavier elements. Ni (Z=28) and many lighter elements have abundances which are some three to six orders of magnitude lower than the abundance of H. Copper (Z=29) and heavier elements in the periodic table are 7 or more orders of magnitude less abundant than H. (Details on FIP and $\log_{10}$ of abundances for the more prominent elements in the sun are given in Table 1). Therefore, when dealing with spectroscopic properties of the upper solar atmosphere (T_e > 1×10^4 K), ions from elements with Z > 28 are seldom detected.

Table 1

Element		I.P. (eV)	Photospheric Abundances[1]	Ave. Coronal Abundances	Extreme[4] Abundances
1	H	13.6	12.00	12.00	12.00
2	He	24.6	10.99	10.99	10.99
6	C	11.3	8.56	8.56	8.56
7	N	14.5	8.05	8.05	8.05
8	O	13.6	8.93	8.93	8.93
10	Ne	21.7	8.11[2]	8.11	8.11
11	Na	5.2	6.33	6.87	7.53
12	Mg	7.6	7.58	8.12	8.78
13	Al	6.0	6.47	7.01	7.67
14	Si	8.2	7.55	8.09	7.75
16	S	10.4	7.21	7.61	8.00
18	Ar	15.8	6.65[2]	6.65	6.65
20	Ca	6.1	6.36	6.90	7.56
26	Fe	7.9	7.51[3]	8.05	8.71
28	Ni	7.6	6.25	6.79	7.45

[1] Adopted from Grevesse and Anders (1989). Abundances are given in units of $\log_{10}$, and $\log_{10}$ H is fixed at 12.
[2] Average of prediction by Grevesse and Anders (1989) and impulsive flare measurement by Feldman and Widing (1990).
[3] The solar-system abundance of Fe based on meteorites was used. For justification see McKenzie and Feldman (1991).
[4] Predictions of the extreme abundances expected to be found in open magnetic coronal field structures.

The main method used in determining elemental abundances is based on the comparison of line ratios from ions emitted over similar temperature ranges.

The intensity of an optically thin spectral line (I_{ij}) between the levels i and j is given by:

$$I_{ij} = C\, P_{ij}\, A_{el}\, \langle G(T)\rangle \int n_e^2 dV$$

where C is a constant that includes geometrical factors, P_{ij} is an atomic quantity which is dependent upon the particular transition, A_{el} is the elemental abundance relative to H, $G(T_e)$ is a quantity that describes the relative fractional abundance of the particular ion, and the integral of $n_e^2 dV$, commonly called the Emission-Measure (EM), describes the product of the amount of plasma in a particular region multiplied by the electron density. For simple resonance lines and using reasonable assumptions of solar plasma properties, P_{ij} and $G(T_e)$ can be calculated with an accuracy of ≤ 20%. A_{el} and EM are then the only quantities left to be determined. The EM is a plasma property independent of the abundance of the elements. Therefore, in comparing line intensities emitted by ions from different elements, the two quantities can be evaluated independently. In general, when lines from several ions of the same element are present, products of EM and A_{el} are plotted against temperature. By adjusting the A_{el}, EM product, plots from other elements are matched to fit the original plot.

III. A CORONAL ENRICHMENT MODEL BY LOW FIP ELEMENTS

I will first start by presenting a model that, at least qualitatively, can explain modifications of coronal abundances, and later I will show that the findings are compatible with the proposed model.

In the solar atmosphere, where the temperature is $T_e <$ 1x10^4 K, some elements stay neutral while others become ionized. The state of the atom is determined by the FIP. An atom stays neutral or becomes ionized depending on the atom's ionization rates and the ion's recombination rates (see Fig. 2). At $T_e \approx$ 1x10^4 K elements with I.P. <10 eV are ionized, elements with I.P. >11 eV are mostly neutral, and S (I.P.=10.4 eV) is at an intermediate stage.

The central assumption of the model is the presence of a strong electromagnetic field (**E**) that moves all the ionized atoms from the lower chromosphere into the region where all atoms are already ionized. The amount of diffusion depends on the strengths and orientations of the **E** field and the local magnetic field (**B**). The I.P. which determines the dividing line between the two FIP groups depends on the local temperature of the lower chromosphere. If the **E** and **B** fields are aligned and oriented in an approximately vertical direction, the ions tend to move into the corona with little resistance. In cases where fields are perpendicular to each other the resistance becomes high. The abundance of the coronal structure is determined by the amount of plasma already present in the coronal structure, the relative amount of un-processed photospheric material that moves into the structure, the coronal material that escapes the structure, and the diffusion of "low" FIP ions from the lower chromosphere into the structure. Examples of abundance enrichments occurring under various solar conditions, which seem to be consistent with the proposed model, are given below.

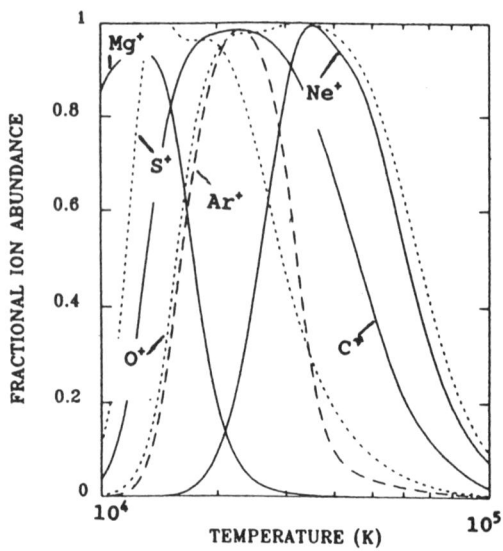

Fig. 2 Fractional abundances for some prominent ions

IV. VARIATIONS IN CORONAL ABUNDANCES RESULTING FROM CHANGES IN THE LOWER CHROMOSPHERE HEATING

a) The Average Solar Atmosphere

Solar plasmas at temperatures of $1\times10^5 < T_e < 1\times10^6$ K are concentrated primarily in structures which are, on the average, several thousand km high. Resonance lines emitted from these plasmas are mostly optically thin. Therefore, monochromatic images of the sun in these lines are marked by limb brightening rings (See The "Atlas of Extreme Ultraviolet Spectroheliograms From 170 - 625 Å" by Feldman et al. (1987)). Average coronal abundances can be determined from intensity ratios of such rings. Although, so far, no detailed abundance determination has been made from ring intensities, a preliminary estimate shows that, indeed, coronal abundances are different from photospheric abundances by a factor 3-4, as suggested by SW and SEP measurements. Meyer (1985) showed abundance differences of

3-4 between the average corona and photosphere.

b) Un-Processed Photospheric Material In The Corona
During the Skylab mission approximately a dozen eruptive prominences were observed in the 300 - 625 Å wavelength range with the S082-a spectroheliograph. The 1974 January 17 event was the only eruptive limb occurrence to be visible, beside the He I 584 Å and He II 304 Å images, in many intense and hot emission lines $T_e \leq 1 \times 10^6$ K.

Intensities of bright images of O III-IV, Ne III-VII, Mg VI-IX, and others were measured. For each of the elements an EM curve was derived, assuming a set of elemental abundances. In order to obtain a general emission measure curve that agrees with the individual EM curves, elemental abundances were adjusted relative to Mg (Fig. 3). Perhaps the most interesting result obtained, was the O/Mg abundance ratio. The O/Mg abundance ratio in the average corona is 6, while the photospheric ratio is 21.7 (Table 1). The ratio obtained from the 1974 January 17 event was 17.8 (Widing et al. 1986). The closeness of the ratio to the photospheric value is an indication of an expulsion into the corona of "raw" photospheric material.

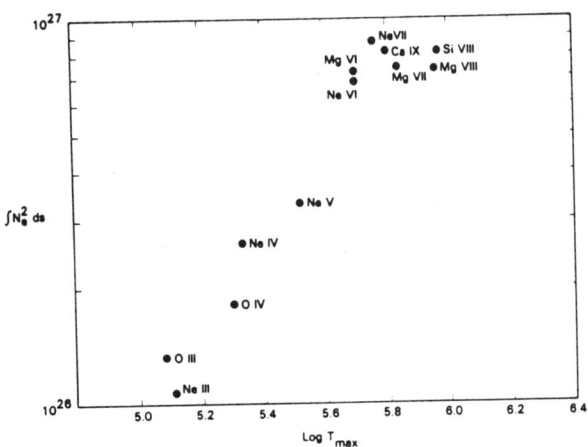

Fig. 3 Adjusted EM curve for the 1974 January 17 event

c) The Lower Chromosphere Under Extreme Heating Conditions
Spectra from the 1973 December 2 flare ($N_e = 2 \times 10^{11}$ cm^{-3}) were also recorded in the 300-625 Å wavelength range by the S082-a spectroheliograph. (See Vol. II of The "Atlas of Extreme Ultraviolet Spectroheliograms From 170 60 625 Å" by Feldman et al(1987)). Line intensities emitted by the flare plasma were measured for ions of O III-IV, Ne II-VII, Mg VI-IX, and Ar V-VII (Feldman and Widing 1990). An O/Mg abundance ratio of 22 was determined, thus identifying the flare plasma source as photospheric.

Abundances of Ne and Ar in the photosphere where never before determined, since resonance lines of their neutral atoms are too energetic to be excited under photospheric conditions. Theoretical considerations and measurements from astrophysical objects outside the solar system placed their abundance relative to Mg as Ne/Mg=3.1 and Ar/Mg=0.10. Abundance ratios in this flare were found to be Ne/Mg=3.6, and Ar/Mg=0.15, in good agreement with the predictions.

Flare plasmas are probably confined to small loops by strong **B** fields. This particular flare can be looked upon as an example of a sudden heating ($T_e \gg 21.7$ eV, the FIP of Ne) taking place in unprocessed layers of the solar atmosphere, causing an instantaneous ionization of all the neutrals without regard to their FIP. As such the dividing line between the FIP groups moved to include all elements in the low FIP group, thus causing coronal and photospheric abundances to become identical.

c) <u>The Lower chromosphere under intermediate heating Conditions</u>

McKenzie and Feldman (1991) have studied the abundance ratios in active regions and flares between Fe/Mg, Fe/Ne, Fe/O and O/Ne. They found that the relative abundances of the two low FIP elements, Fe and Mg, which differ considerably in atomic weights, did not change. However, the Fe/Ne ratio varied by a factor of 4. The Fe/O abundance ratio showed a much smaller variation, and the O/Ne abundance ratio varied by as much as a factor of two (Fig. 4). McKenzie and Feldman (1991) interpreted the O/Ne variation as an indication that the FIP dividing line shifted toward higher values in some of the observations. Under intense heating of the lower chromosphere, elements with FIP of 13 eV may become "low" FIP, causing the observed effect.

V. VARIATIONS ON RELATIONSHIPS BETWEEN THE E AND B FIELDS

a) <u>Coronal **E** and **B** Field In The Same Orientation</u>

Polar plumes are low density (2×10^9 cm^{-3}) open magnetic field structures. Their plasma temperature is typically $3 \times 10^5 < T_e < 1 \times 10^6$ K. A bright plume was observed by the SO82-a instrument in lines of Ne VI-VII, Mg VI-IX and Ca IX-X. Widing and Feldman (1989,1991) used the EM technique to find that the Mg/Ne abundance ratio is a factor of 10 larger than the photospheric ratio. However, the abundance ratio of Mg/Ca, two low FIP elements, did not differ from the expected value.

The polar plume is an example of vertically oriented **E** and **B** fields. The SO82-a data shows additional examples of open magnetic field structures in which ratios between low and high FIP elements are as large as 15.

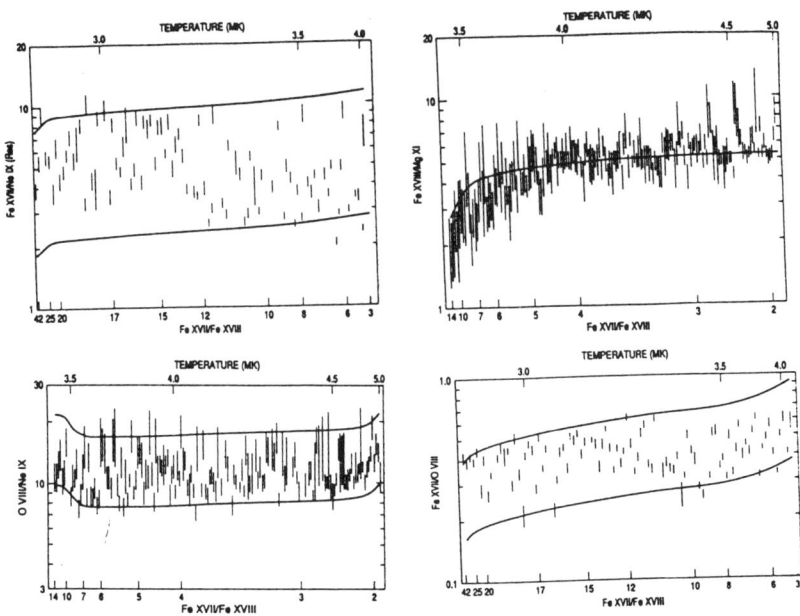

Fig. 4 Abundance ratios of O, Ne, Mg and Fe in flares b)

b) **An E Field Perpendicular To A Strong B Field**

The NRL HRTS spectrograph (Brueckner et al. 1977) imaged a cord along the solar surface, ≈1000" long, and ≈1" wide. The spatial resolution along the slit was ≈1". The wide spectral coverage of the instrument (1200 - 1700 Å) provided a detailed, simultaneous look into properties of the atmosphere from the lower chromosphere to the top of the lower transition zone, in each of 1000 slit elements.

During the 1975 July 21 flight the slit was positioned along a narrow line extending from about the solar center to the limb. The slit crossed the center of a sunspot and the surrounding plage area (Dere et al. 1982). An investigation of the sunspot line intensities relative to intensities of lines from surrounding plage and quiet sun areas revealed unusual behavior. Fig. 5 is a composite of slit images produced from a 51 s exposure of the HRTS spectra. The images show the sunspot and a nearby plage area in lines from Si II, C II, Si III, Si IV, C III, He II, S IV, C IV, O IV, N V, and O V. Two marks placed on the sides of the figure point to the sunspot (left mark) and to a bright plage area (right mark), respectively. Notice that in the C II, C III, and C IV lines the relative intensities between the plage and sunspot are ≤ 1. In the lines of Si II, Si III, and Si IV, however, the relative intensities between the plage and sunspot are >> 1. The intensity ratios between the plage

178 The Upper Solar Atmosphere

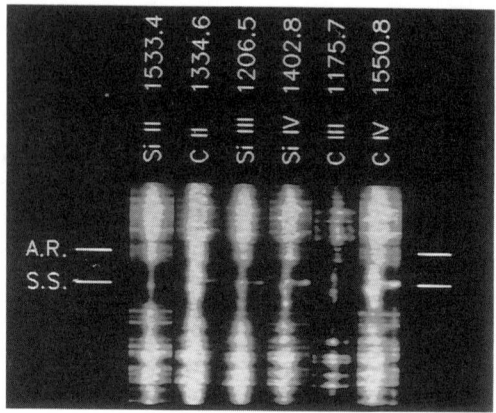

Fig. 5 A composite of slit images from the HRTS instrument.

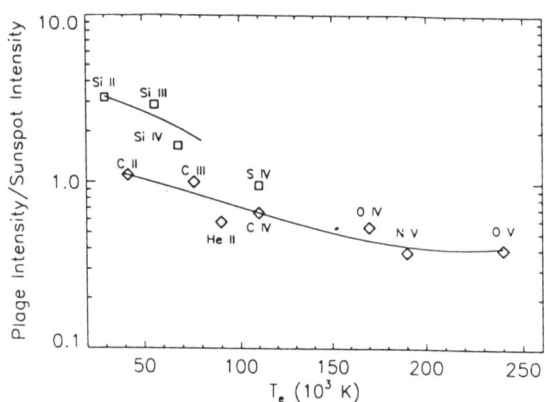

Fig 6. Intensity ratios of the sunspot and plage regions.

and the sunspot are plotted as a function of temperature in fig. 6. The fact that all three Si ions behave similarly to each other and differently from the three C ions is an indication of changes in the Si/C abundances between the two regions.

Dere et al. (1982) showed that the Si IV / C IV intensity ratio in the sunspot is reduced by a factor of ≈ 3 relative to similar values in nearby active regions, an indication of photospheric abundance in the sunspot plasma.

According to the proposed model the strong horizontal sun spot **B** field traps the chromospheric ions, preventing the **E** field from pulling them into the corona; hence hindering the enrichment of the corona by low FIP ions.

c) <u>Typical Cases Of Abundance Variations In Coronal Loops</u>

S082-a images of bright coronal loops in the 400 Å lines of Ne VI and Mg VI are excellent indicators of the abundance ratios between high and low FIP elements within these structures. A close investigation of such images (i.e. Vol. IV of "Atlas of Extreme Ultraviolet Spectroheliograms From 170 to 625 Å" by Feldman et al. (1987)) reveals Ne/Mg abundance ratios that span most of the range from photospheric to extreme coronal. It is believed that these ratios are caused by variations in strength and orientation of the electromagnetic field, combined with variations of the lower chromosphere heating conditions.

VI. ARE LOW FIP ELEMENTS ENHANCED OR HIGH FIP ELEMENTS REDUCED IN THE UPPER SOLAR ATMOSPHERE

A basic question to answer is, what changes? Are abundances of high FIP elements reduced, or low FIP elements enhanced, in the transition from the photosphere to the corona? Elemental abundances are measured relative to hydrogen, therefore, a simultaneous determination of hydrogen and the other elemental abundances is necessary. At upper solar atmosphere temperatures no hydrogen lines are visible. Veck and Parkinson (1981) tried to overcome this difficulty by comparing the intensity of spectral lines from highly ionized elements with the intensity of the continuum, which, in part, depends on the H abundance. They concluded that high FIP elements are reduced in the corona.

Feldman et al (1990), used relative intensities between the sunspot and the plage to answer the question. Fig. 7 is a composite images from a HRTS 0.34 s exposure showing the same sunspot as in Fig. 6, but a different plage area marked "A.R." The images shown are O I 1306 Å, H I Ly α 1216 Å, Si II 1533 Å, C II 1335 Å, and Si III 1206 Å. The C II intensity ratios between the A.R. and the sunspot resemble the neutral O I and H I Ly α images, while the Si II and Si III images are different, an indication that low FIP elements are enhanced in the corona.

Moreover, in open field regions such as polar plumes, low FIP, low abundance elements like Na become intense, while in regions of photospheric abundances they are too faint to be visible, again an indication that the low FIP elements are enhanced in the corona.

VII. CONCLUSION

It was shown that the coronal vs. photospheric abundance ratios can vary by an order of magnitude and perhaps even more (Table 1). Increases in coronal abundances of low FIP elements results from low FIP ions formed in regions of the low chromosphere diffusing into coronal regions.

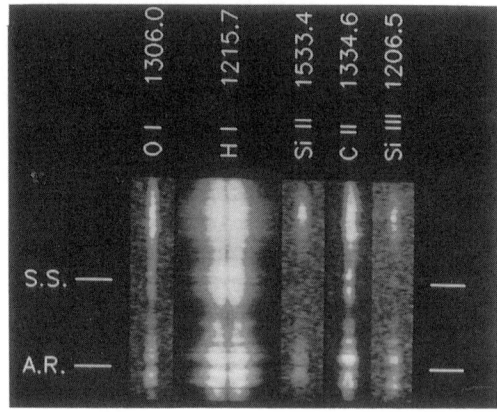

Fig. 7 A composite of slit images from a 0.34 s exposure.

REFERENCES

Arnaud, M., and Rothenflug, R. 1985, Astr.Ap. Suppl., **7**, 291.
Breneman, G.H., and Stone, E. C. 1985, Ap.J. (Letters), **299**, L57.
Brueckner, G.E., Bartoe, J.- D.F., VanHoosier M.E., 1977, in Proc. OSO 8 Workshop: Univ. of Colorado Press, p.380.
Coplan, M.A., Ogilvie K.W., Bochsler, P. and Geiss, J. 1990, Solar Phys. **128**, 195.
Dere, K.P., Bartoe, J.-D.F., Brueckner, G.E. 1982, Ap.J., **259**, 366.
Feldman, U., Purcell, J.D., and Dohne, B.C. 1987, An Atlas of extreme ultraviolet specrtoheliograms from 170 to 625 Å (NRL Report 90_4100 and 91_4100) (Washington: Naval Research Laboratory).
Feldman, U., and Widing, K.G. 1990, Ap.J., **363**, 292.
Feldman, U., Widing, K.G., and Lund, P.A. 1990, Ap.J. (Letters),**364**, L21.
Gloeckler, C., and Geiss, J. 1989, in AIP Conf. Proc. No. 183 Cosmic Abundances of Matter, ed. C. J. Waddington (New York:AIP,p. 49).
Grevesse and Anders (1989)in AIP Conf. Proc. No. 183 Cosmic Abundances of Matter, ed. C. J. Waddington (New York : AIP, p. 1).
McKenzie, D.L., and Feldman, U. 1991, Ap.J., Submitted
Meyer, J.-P. 1985, Ap.J. Suppl. **57**, 173.
Veck, N.J., and Parkinson, J.H. 1981, MNRAS, **197**, 41.
Widing, K.G., and Feldman, U. 1989, Ap.J., **334**, 1046.
Widing, K.G., and Feldman, U. 1991, Ap.J., Submitted
Widing, K.G., Feldman, U., and Bhatia, A.K. 1986, Ap.J., **308, 982.**

THE FE L-SHELL SPECTRUM IN
COMPACT ASTROPHYSICAL X-RAY SOURCES

D.A. Liedahl and S.M. Kahn
Physics Department and Space Sciences Laboratory
University of California, Berkeley, CA 94720

W.H. Goldstein and A.L. Osterheld
Lawrence Livermore National Laboratory,
P.O. Box 808, Livermore, CA 94550

ABSTRACT

Compact X-ray sources derive their luminosities from the conversion of gravitational potential energy through accretion. A centrally-produced hard X-ray continuum photoionizes the surrounding accretion flow, which can produce discrete X-ray emission. To date, as applied to compact X-ray sources, a detailed diagnostic approach to the analysis of spectroscopic data has been largely ineffective, owing to the limited quality of available line spectra and the lack of appropriate models of X-ray line emission in the presence of an ionizing continuum radiation field. We are involved in an atomic modeling program to investigate the discrete spectral response to variations in ambient plasma conditions, with the aim of establishing a new set of plasma diagnostics appropriate to this physical regime. We discuss the physical mechanisms and applications of several of the diagnostics which have been discovered in the Fe L-shell spectrum, under conditions appropriate to X-ray photoionized plasmas.

I. INTRODUCTION

The generation of radiation through accretion is one of the dominant light-producing mechanisms in the Universe. The integrated light output of quasars, which are believed to be powered by accretion onto supermassive black holes, rivals that produced through stellar fusion. This is all the more remarkable given that a quasar can outshine a typical galaxy by several orders of magnitude ($L = 10^{41} - 10^{47}$ erg s^{-1}) while occupying a negligible fraction of its volume (see the article by J. Weisheit, this volume). The origin of the inflowing matter is the host galaxy in which the AGN resides. Quasars and other active galactic nuclei (AGN) are known to be strong X-ray sources[1].

Within our own galaxy, binary systems containing compact stellar objects (white dwarfs, neutron stars, or black holes) constitute some of the brightest X-ray sources in the sky. For example, a typical neutron star binary X-ray source can radiate, in X rays, up to 10^5 times the Sun's bolometric luminosity. The binary X-ray sources are powered by the transfer of matter within the system, from a main-sequence or evolved star to the compact star. There are a handful of these objects in which the compact star is a black hole "candidate", the most famous of which is Cygnus X-1. Observations of objects such as Cyg X-1 are currently the best means by which to study black holes. Also, to some extent, a black hole binary X-ray source may represent a scaled-down version of an AGN, in spite of the dissimilarities in the origin of the accreted mass. The neutron star binaries exhibit widely varying phenomena, such as X-ray pulsars, X-ray bursts, quasi-periodic oscillations, hard X-ray cyclotron lines, *etc.*, and involve exotic physical scenarios, such as complex MHD, relativistic effects (special *and* general), the interaction of highly-ionized matter with high-energy radiation fields, relativistic shocks, and so on.

Astrophysical models of accretion-powered sources which involve these complex processes often do not provide a distinct observational signature. By contrast, discrete spectra are sensitive to several parameters and vary in ways which can usually be calculated to degrees of accuracy sufficient to provide constraints on the models. X-ray spectroscopy, in particular, can be a valuable probe of the high temperature and/or highly ionized regions of compact X-ray sources. Spectroscopic data from previous X-ray observatory missions has been analyzed in the context of coronal plasma emission[2]. Convenient codes and compilations exist for this purpose[3,4]. However, the assumptions which underlie these codes are likely to break down in the circumsource medium of a compact X-ray source (see Part III). The use of coronal plasma emission codes has led to apparent inconsistencies in the interpretation of discrete X-ray data from X-ray binaries[5].

Owing to recent advancements in atomic modeling techniques[6] and the growing interest in the physics of multiply-charged ions, largely motivated by laboratory applications, it is now possible to efficiently address some of the atomic physics issues related to accretion-powered X-ray sources. This development is timely. Two "facility-class" X-ray observatories, the Advanced X-ray Astrophysics Facility (AXAF)[7] and the X-ray Multi-Mirror Mission (XMM)[8], are scheduled for launch by the late 1990s. The spectroscopic instruments being developed for these missions will achieve order-of-magnitude improvement in both spectral resolution and detector area compared to previous extrasolar missions.

We have been involved with the re-calculation of X-ray line spectra, emphasizing atomic kinetics issues, in an attempt to discover new discrete spectral diagnostics relevant to X-ray photoionized plasmas, which are likely to exist in the circumsource medium of an accretion-powered source. In this paper we present an overview of the results of this work. In Part II, to provide an astrophysical context for our work, we present a brief description of the physics of accretion-powered X-ray sources. Part III is a discussion of microphysical processes relevant to astrophysical plasmas. We highlight the differences between line formation processes in coronal plasmas and photoionized plasmas. A discrete spectral signature of X-ray photoionization is presented in Part IV. In Parts V and VI, we discuss the unique diagnostics of electron density and temperature, respectively.

II. ACCRETION-POWERED X-RAY SOURCES

We discuss the topic of accretion-powered X-ray sources mainly in terms of X-ray binaries (XRBs), since there are, as yet, no published discrete soft X-ray (5-45 Å) spectra from AGNs or cataclysmic variables. Two observatory missions, *Einstein* and *EXOSAT*, resulted in the acquisition of moderate-resolution ($E/\Delta E$ = 10-50) spectra from X-ray binaries. However, the microphysical scenario which we establish should apply to all classes of accretion-powered sources.

There are two classes of XRB, low-mass[9] and high-mass[10], characterized by the mass of the companion. Binary systems with companion masses of about one solar mass (1 M_s) or less are referred to as low-mass systems, while high-mass XRBs usually contain young O or B stars with masses of about 10 M_s or more. To first order, this classification scheme correlates with the mechanism by which mass is transferred from the companion to the compact star.

In a high-mass XRB, the O or B star undergoes significant mass loss through a stellar wind, driven by radiation pressure on resonance lines of ions in the wind. Mass loss rates can be as high as 10^{-5} M_s yr^{-1}, a fraction of which can be captured by the orbiting compact star. Mass transfer in a low-mass XRB proceeds by Roche lobe overflow. The Roche lobe is the gravitational equipotential surface which intersects the gravitational neutral point between the two stars. At later stages in the evolution of the companion, its surface expands to fill the Roche lobe, spilling matter through a region near the neutral point. The transferred matter, which may be subsequently captured by the compact star, has angular momentum with respect to the compact star, owing to the binary orbital motion. Thus, matter which is captured is forced to orbit the compact star, forming an accretion disk in the binary orbital plane. Binary separations (center-to-center) range from about 1.5 x 10^{10} cm to a few x 10^{12} cm. A typical accretion disk has a radius of about 10^{11} cm. Lying at distances of 1-10 kpc (1 kpc = 3.1 x 10^{21} cm), the angular sizes of XRBs are too small to allow imaging. Again, this points out the necessity for spectroscopic studies.

The extraction of gravitational potential energy from disk material requires the dissipation of angular momentum. This is accomplished by viscous stresses in the disk. These stresses also dissipate energy, which heats the disk material, resulting in radiation. As matter slowly spirals down the potential well, half of its initial energy is radiated. If the compact star is a neutron star, the remaining kinetic energy must be converted through a process which allows the matter to come to rest on the stellar surface. Shocks are often invoked to account for this conversion of kinetic energy to radiation. The total X-ray luminosity produced through this process is given roughly by $L_X = GMR^{-1} \, dM/dt$ where M and R are the mass and radius of the compact star, G is the gravitational constant, and dM/dt is the mass capture rate. (If the compact star is a black hole, R is usually taken to be the radius of the last stable circular orbit.) The release of this energy in a volume comparable to a neutron star explains the high-energy nature of compact X-ray sources. In spite of the success of this simple picture in explaining the gross features of XRB phenomena, our level of understanding is rather superficial, with many details yet to be worked out. For a review of accretion physics, see Ref. [11].

III. ATOMIC PHYSICS AND X-RAY SPECTRA

We define the soft X-ray band as the wavelength region 5-45 Å. This band contains the K-shell transitions of the cosmically abundant elements C, N, O, Ne, Mg, Si, and S, and the L-shell transitions of Fe and Ni. Of particular importance is the Fe L-shell spectrum, that is, the spectrum involving transitions terminating on the $n = 2$ shell in Li-like through Ne-like Fe (Fe XVII-XXIV). The Fe L-shell ions dominate the discrete X-ray spectrum from astrophysical sources over the 5-18 Å wavelength band, including the solar corona[12].

It is certain that most of the discrete soft X-ray emission in XRB spectra is Fe L-shell emission. However, the pattern of line ratios in the spectra do not conform to those observed in, for example, the solar corona. In the early analyses, even line identifications were, at best, tentative. It was suggested that the discrepancy arose simply because the plasma emission codes and line lists are invalid in the X-ray emission line region (XELR) of accretion-powered X-ray sources[5].

The "coronal assumption" consists of modeling the X-ray line emission as being produced by an equilibrium plasma in which the ionization and level populations are functions only of the electron temperature. In addition, one assumes that the electron densities are low enough that excited state populations are negligible relative to the ground state population. Finally, the line optical depths are assumed negligible. Line emissivities, in coronal equilibrium, are functions of a single parameter, the electron temperature. To first order, the more intense X-ray lines from coronal plasmas are those whose upper levels are rapidly populated by collisional excitation from the ground state.

The assumption of coronal equilibrium, the fundamental assumption used in assembling the common plasma emission codes, breaks down if photoionization is important. Static models of plasmas surrounding compact X-ray sources, called X-ray nebular models[13,14], have been calculated in order to investigate the effects of a hard X-ray ionizing continuum on the state of the gas. We employ the term, X-ray photoionized nebula (XPN), to refer to such a plasma in an astrophysical setting. The state of the nebular gas is characterized by the 'ionization parameter', $\xi = L/nr^2$, where L, n, and r are the ionizing luminosity, gas density, and distance from the central X-ray source, respectively. That is, the charge state distribution and temperature depend only on ξ in the optically thin limit.

Calculations of X-ray nebular models have emphasized the energetics and ionization balance over the details of line formation. Thus, the models fall short of providing testable predictions concerning the X-ray line spectrum. However, schematic treatments of the trace ion (e.g., Fe species) physics are not likely to appreciably affect the predicted charge state distribution and temperature structure. Our goal has been to investigate the details of line formation, given the conditions predicted by X-ray nebular models.

For our purposes, the crucial result from X-ray nebular models is that the X-ray-emitting ions exist at temperatures much lower than in coronal equilibrium. For example, the peak abundance of Fe XVII (Ne-like) occurs at about 2×10^6 K under coronal conditions[15], while, under XPN conditions, the corresponding temperature is only about 10^5 K [14]. This is an indication of the degree to which photoionization dominates the atomic processes in XPN. This relatively low XPN temperature means that the efficiency of collisional excitation of X-ray-emitting levels is drastically reduced. In effect, the number of electrons in the Maxwellian electron distribution with sufficient energy to induce a $\Delta n > 0$ transition is negligible, even at the temperature of peak ionic abundance. This fact is embodied in the Boltzmann factor, $e^{-\Delta E/kT}$, for the transition. In XPN, the argument of the exponential is on the order of 100, while, in coronal equilibrium, arguments of about 1-5 are relevant.

Radiative recombination (RR) is a decreasing function of temperature and replaces collisional excitation as the dominant excitation mechanism in XPN. Line emissivities are determined by a completely different set of cross-sections and by the radiative cascades which follow recombination.

In order to model X-ray spectra under a variety of conditions, we have employed the HULLAC atomic physics package[16,17]. Atomic structures for the ions Fe XVI-XXIV (Na-like through Li-like) are calculated in the using a relativistic, multi-configuration, parametric potential technique. Radiative rates include E1, M1, E2, and M2 moments. Cross-sections for electron impact excitation are calculated using the quasi-relativistic distorted wave approximation. Recombination cross-sections are calculated from the Milne relation using the corresponding photoionization cross-sections[18,19]. Rate coefficients for collisional excitation and RR are obtained by integrating the cross-sections over a Maxwellian electron velocity distribution. Using a fixed charge state distribution, the rate equations are solved for the level populations, from which the emissivity spectrum is determined. We do not consider radiative transfer. A typical atomic model includes several hundred levels through $n = 4$ and all the rates which connect them.

IV. DISCRETE SIGNATURE OF PHOTOIONIZATION

We have argued that the XELR may be photoionized and now discuss the spectroscopic consequences of this possibility. It is not obvious, *a priori*, that a substantially distinct pattern of line ratios should arise. However, the multielectron Fe L-shell ions have excited-state configurations with several fine-structure levels. Parity and angular momentum selection rules dictate the cascade routes for recombining electrons and lead to spectra which differ dramatically from the coronal case. This is demonstrated in Figure 1 for Fe XVI-XIX [20].

The most important conclusion to draw from Figure 1 is that the differences in the line spectra are obvious, even at a qualitative level. Photoionization clearly manifests itself, thus establishing the XPN regime as the proper context in which to identify lines and interpret quantitative diagnostics. Failure to make this distinction can lead to line misidentification, which then leads to incorrect determinations of the physical conditions in the XELR. To illustrate, consider the 15 Å line in the model coronal spectrum of Figure 1. This line is a $3d-2p$ transition in Fe XVII and is one of the "benchmark" lines in X-ray astronomy, commonly identified in the solar X-ray spectrum[12]. Thus, it is natural to associate this feature with Fe XVII in *any* X-ray spectrum, since, according to existing line lists, there are no other lines near 15 Å with comparable intensities. Furthermore, this identification establishes a coronal (or collision-driven) context for line formation and an electron temperature of a few million degrees, corresponding to the Fe XVII ionization temperature.

The model XPN spectrum, calculated with $kT = 10$ eV, shows that a strong 15 Å feature is expected from Fe XIX, in this case a $3s-2p$ transition. XRB spectra often show features at 15, 16, and 17 Å, consistent with the model XPN spectrum. However, the strong 16 Å line is *not* consistent with coronal conditions[5]. Thus, we believe that the XELRs are photoionized, at least in some of the cases so far studied, and that previous analyses of XRB spectra have resulted in a severe overestimate of the temperature.

To summarize, the spectral signatures of X-ray photoionization are large *(3s-2p)/(3d-2p)* ratios in the Fe L-shell ions, indicating recombination-dominated population kinetics. The strong $3d$ lines in coronal spectra are collision-driven and, therefore, are weak in the XPN spectrum. The $2p^k 3s$ ($1s^2 2s^2$ core implied) configurations are favored in this regime simply because of their status as the energetically lowest-lying among configurations with an electron excited to the M shell. Most of the levels within the higher-lying configurations, $2p^k 3d$ and $2p^k 3p$, depopulate through the various $2p^k 3s$ states, thus explaining the shift in $3d$ lines to $3s$ lines. The effect is most pronounced for the charge states near Ne-like (Fe XVII) but persists throughout the series.

The Lα energy centroids are plotted vs. charge state in Figure 2. The centroids are shifted downward in energy by 10-50 eV ($\Delta\lambda = 0.10 - 1.20$ Å). The energy centroid is defined by $\langle E \rangle = \Sigma j_i E_i / \Sigma j_i$ where j is the photon emissivity and the sum is over all 3-2 lines for a given charge state.

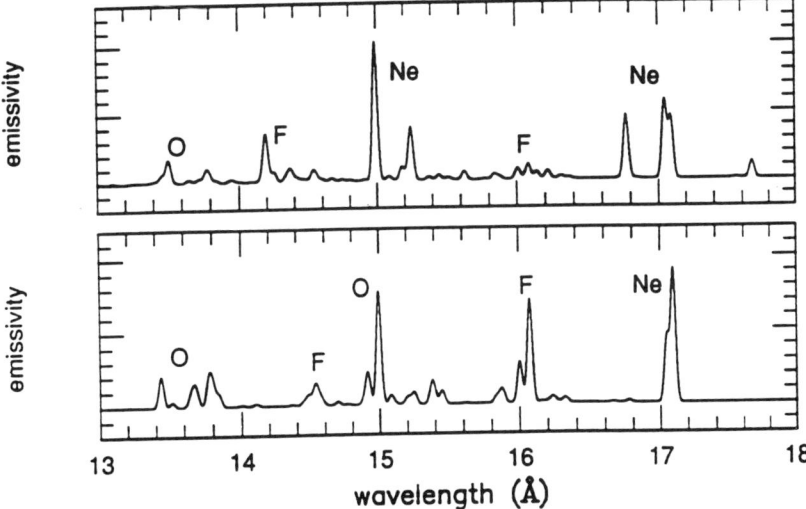

Figure 1. Comparison of model emission rate spectra for Fe XVI-XIX (labeled by isosequence) in two environments. *Top:* Coronal spectrum at 500 eV. The intense lines between 13 and 15.5 Å are $3d-2p$ transitions, while those at longer wavelengths are $3s-2p$. *Bottom:* XPN spectrum at 10 eV. Intense lines are $3s-2p$ transitions, including the 15 Å Fe XIX feature. Resolution in both panels is $\Delta\lambda = 0.05$ Å. The emissivity scales are in arbitrary units.

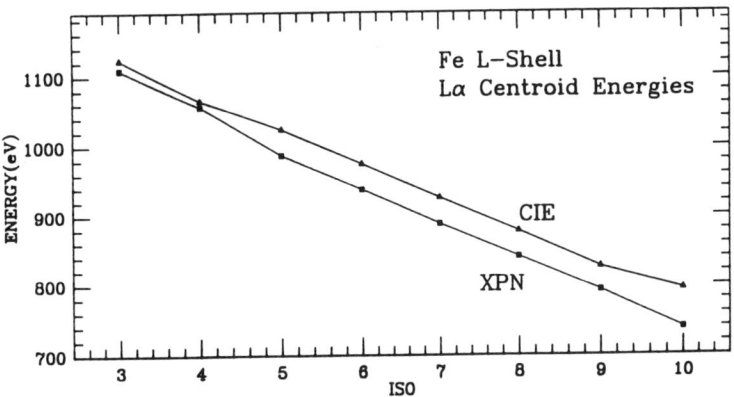

Figure 2. Centroid energies of Lα transitions in Fe L-shell ions plotted against isoelectronic sequence (*e.g.*, iso 3 corresponds to Li-like Fe XXIV) in coronal ionization equilibrium (solid line) and an XPN (dotted line).

V. DENSITY DIAGNOSTICS

Mass accretion rates for high-L_x sources range from 10^{16} - 10^{17} g s^{-1}. The radial flow velocity depends, among other factors, on the viscosity mechanism, which is poorly understood. Therefore the density distribution and total mass of the accretion flow are not known. One of the important tasks of X-ray spectroscopy is to provide model-independent density diagnostics. For the range of densities expected to exist in accretion-powered sources[21], the Fe L-shell ions provide an excellent set of density-sensitive line ratios, which may help to constrain some of these unknowns.

Suppose the density of some particular ionization zone is determined spectroscopically. Then, by invoking X-ray nebular calculations, the relation of ξ and n can be used to deduce r, the distance of the XELR from the central source. The volume emission measure for a given ionization zone, $EM_{ion} = n^2 V_{ion}$, determined from the measured line flux, can then be used to constrain models of the source geometry, for example, by relating r and V_{ion}.

We have investigated the density dependence of Fe L-shell spectra at XPN temperatures over the range of electron densities, 10^{11}-10^{16} cm^{-3}, and have found several useful line ratios for the ions Fe XVII-XXI [21]. The Fe L-shell ions have several low-lying metastable states, formed by the variety of angular momentum couplings of $2p_{1/2}$ and $2p_{3/2}$ electrons. These low-lying excited states have energy separations of order 10 eV and are easily populated by collisional excitation. Also, they are metastable to radiative decay. As the electron density increases, the metastable state populations increase toward their LTE values, which can be an appreciable fraction of the ground state population. Subsequent recombination of the metastable state begins to compete with recombination of the ground state. Schematically, a recombination can be represented by *(core)* -> *(core) nlj*. For example, Fe XVIII has two cores that can appreciably influence the population

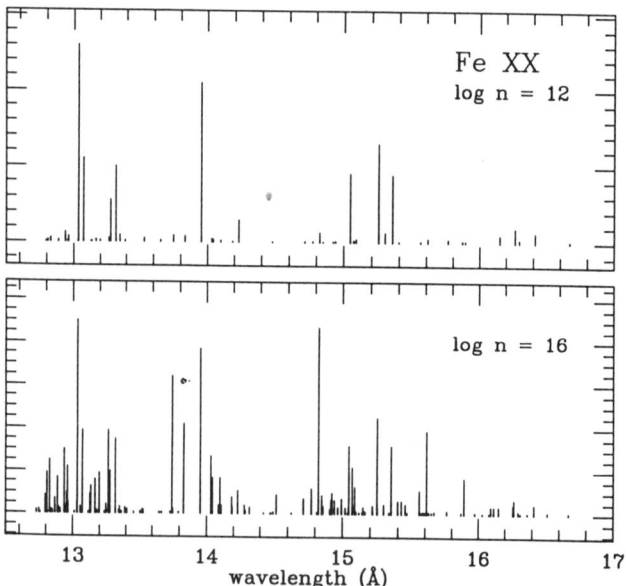

<u>Figure 3</u>: N-like Fe XX line power spectra at two densities, 10^{12} and 10^{16} cm^{-3}, with $kT = 10$ eV, a representative XPN temperature. All transitions are Lα. The complexity is attributable to the ease with which C-like Fe XXI metastable states are populated through collisional excitation at high densities and to the complex Fe XX fine-structure distribution.

kinetics, $2s^2\ 2p_{1/2}^2\ 2p_{3/2}^3$ and $2s^2\ 2p_{1/2}\ 2p_{3/2}^4$. The subsequent cascade "preserves" the core, to the extent that configuration mixing is small. The net effect is the appearance of a distinct set of lines at high electron densities; the complexity of the spectrum increases with density. Line ratios formed by two lines, one from each set, constitute the density-sensitive ratios. Critical electron densities for this process occur over the range 10^{13}-10^{14} cm^{-3}.

The density sensitivity of Fe L-shell ions has been studied for coronal plasmas.[22] (Actually, the coronal approximation begins to break down at these densities. This regime is more properly called 'collision-driven'.) The Fe L-shell critical densities in coronal plasmas are comparable. However, the mechanism is somewhat different in XPN because of the importance of RR. In the coronal case, the metastable states act as sources for $n = 3$ excited states through collisional excitation. Thus, the adjacent charge state plays no role. The coupling of adjacent charge states in XPN results in density-sensitive lines which do not coincide with those which operate in the coronal case.

Figure 3 demonstrates the density sensitivity in an XPN for Fe XX (N-like). The complexity of these spectra manifests the rich Fe XX and Fe XXI fine-structures of the $2s^2 2p^k$ configurations. Of course, the density-sensitive line ratios for a given ion stage are not independent diagnostics, since only a few metastable states in Fe XXI are important sources. However, the redundancy eliminates the problems of line blending with other Fe L-shell lines and K-shell lines of O and Ne species.

VI. TEMPERATURE DIAGNOSTICS

X-ray photoionization creates a unique environment in terms of atomic processes. The XPN electron temperature can be a factor of 20 times cooler than a coronal plasma in a comparable state of ionization. The predominance of RR over collisional excitation in driving the X-ray line emission creates an apparent obstacle to obtaining temperature information from the line spectrum. This difficulty arises from the fact that the X-ray line emissivities have, for practical purposes, roughly the same temperature dependence, $T^{-1/2}$. Furthermore, the presence of some particular charge state, established by identifying an emission line, does not imply a specific narrow temperature range - the 'ionization temperature' severely overestimates the electron temperature.

An upper limit to the temperature of about 100 eV can be inferred by using the simple $3s/3d$ signatures discussed in Part IV. This may be considered a "crossover temperature", above which collisional excitation becomes important. A theoretical lower limit of about 10 eV exists according to X-ray nebular models. This is determined by assuming that the only heating source is photoelectron heating, the minimum heat input required to maintain the ionization level.

Fortunately, the lack of temperature sensitivity of the emission lines in XPN is offset by the increased sensitivity of the L-shell RR continua. The energy width of a Maxwellian electron distribution at temperature, T, is approximately kT. Thus, the energy distribution of photons which comprise the RR continuum is narrow compared to the X-ray energy ($\Delta E/E \ll 1$), where E is about 1-2 keV for transitions from the continuum to the L shell, depending on the charge state being considered. At low temperatures ($kT \ll E$), $\Delta E = kT$ is a very good approximation, thus providing a valuable temperature diagnostic. An example of this effect is shown in Figure 4, which shows the Fe XIX L-shell spectrum at two temperatures. Note that the shape of the Fe XIX (O-like) continuum depends on the temperature of the Fe^{19+} (N-like) ionization zone. Figure 4 demonstrates another feature of RR continua, that the continuum jump decreases faster with temperature than the discrete line-center emissivities. The ratio, (*jump/line peak*) scales roughly as T^{-1}, and serves as a crosscheck on T, as determined from the width. For data with low to moderate spectral resolution, this latter method is likely to be the only one available to determine the temperature.

The Fe L-shell continua lie on the wavelength interval 5-10 Å and, with high resolution spectroscopy, are well separated. These features provide a way to map out the temperature distribution through a succession of ionization zones. Also, by simply identifying such a feature, an upper limit to the temperature may be inferred, since at coronal temperatures, the RR continua

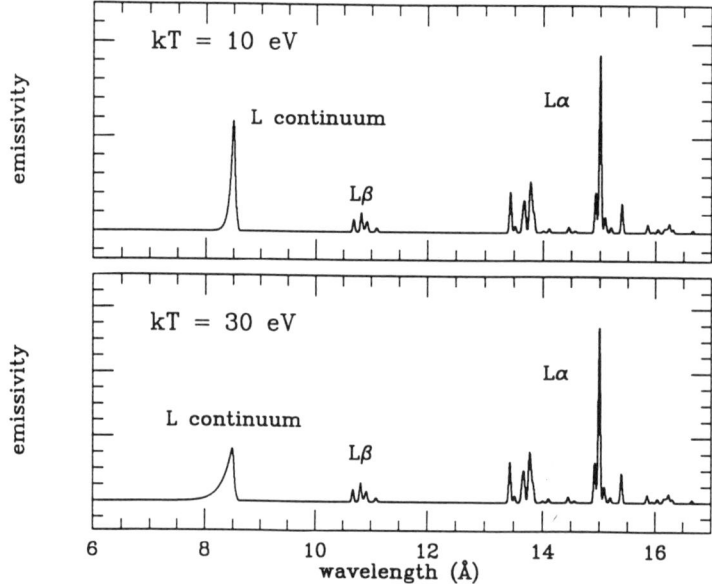

Figure 4. O-like Fe XIX model XPN spectrum at two temperatures, 10 eV and 30 eV. Spectra include Lα, Lβ, and two L-shell RR continua (blended), the latter near 8.5 Å. Comparison of continua at each temperature illustrates the behavior of the width and jump with temperature. Also note the lack of variation of line ratios within Lα and Lβ. Resolution is $\Delta\lambda = 0.05$ Å. The emissivity scales are the same but in arbitrary units.

will be "lost" in the central continuum. Of course, the RR continua of the K-shell species will exhibit similar behavior.

The RR continua may have solved another problem concerning the interpretations of the *Einstein* data. Identifications of strong features which lie above the central continuum spectrum in XRB data[2,5], at wavelengths below 10 Å, have been uncertain. Their intensities have rendered otherwise plausible identifications as unlikely. There are two points in favor of associating these features with RR continua: (1) their positions and intensities, relative to discrete transitions, are consistent with our XPN spectral models and (2) in *Einstein* data from the low-mass XRBs, GX9+9 and 4U 1820-30, Lα transitions between 15 and 17 Å, corresponding to the candidate continuum features, are evident[5]. These findings add further support for the XPN picture in accretion-powered X-ray sources.

VII. SUMMARY

We have shown that the Fe L-shell ions are a versatile class of plasma diagnostics. Up to eleven charge states can emit X-rays over a narrow spectral band, 10-17 Å. The proper interpretation of Fe L-shell spectra can provide information over a wide range, but fine "grid", of physical conditions. Their ubiquity demands a detailed understanding of the atomic kinetics processes leading to X-ray line emission. Here, we have discussed a set of qualitative diagnostics, using the *(3s-2p)/(3d-2p)* line ratios, to discriminate between two distinct environments, a coronal (or collision-driven) plasma and an X-ray photoionized plasma. Recognizing this distinction is a prerequisite to identifying X-ray lines from accretion-powered sources. We have discovered several density diagnostics over the range 10^{12}-10^{16} cm^{-3}, which may be the relevant range for Galactic accretion-powered X-ray sources. We have also shown that electron temperatures, corresponding to each Fe L ionization zone, can be determined according to the widths or jumps of the RR continua.

The launches of *ASTRO-D* (1993), *AXAF* (1997), and *XMM* (1998) will provide unique opportunities to study the physical conditions in a diverse range of X-ray-emitting plasmas. To fully exploit the high quality data expected from these observatories, a versatile set of spectroscopic tools must be developed. The Fe L-shell ions and the He-like ions[23,24] of cosmically abundant elements are the two most important classes of discrete diagnostics in the soft X-ray band. We believe that the Fe L-shell diagnostics discussed here will form an essential part of the array of spectroscopic tools needed to interpret data from accretion-powered X-ray sources.

Work on this project at UC Berkeley was supported by grants from the NASA Astrophysics Data Program, the NASA Long-Term Space Astrophysics Program, and the Institute for Geophysics and Planetary Physics at LLNL. Work at LLNL was performed under the auspices of the U.S. Department of Energy under Contract No. W-7405-ENG-48.

VIII. REFERENCES

1. H. Inoue in *Proc. 23rd ESLAB Symposium on Two Topics in X-ray Astronomy* (ESA SP-296, 1989).
2. S.M. Kahn, F.D. Seward, and T. Chlebowski, *Ap. J.* **283**, 286 (1984).
3. J.C. Raymond and B.W. Smith, *Ap. J. Suppl.* **35**, 419 (1977).
4. R. Mewe, E.H.B.M. Gronenschild, and G.H.J. van den Oord, *Astr. Ap. Suppl.* **62**, 197 (1985).
5. S.D. Vrtilek, D.J. Helfand, J.P. Halpern, S.M. Kahn, and F.D. Seward, *Ap. J.* **308**, 644 (1986).
6. W.H. Goldstein, in *IAU Colloquium 115, High-Resolution X-Ray Spectroscopy of Cosmic Plasmas*, ed. P. Gorenstein and M.V. Zombeck (Cambridge University Press 1990).
7. M.C. Weisskopf, *Space Sci. Rev.* **47**, 47 (1988).
8. A.C. Brinkmann, et al., in *EUV, X-Ray, and Gamma-Ray Instrumentation for Astronomy and Atomic Physics*, ed. C.J. Hailey and O.H.W. Siegmund, *Proc. S.P.I.E.* **1159**, 495 (1989).
9. W.H.G. Lewin and P.C. Joss, in *Accretion-Driven Stellar X-ray Sources*, ed. W.H.G. Lewin and E.P.J. van den Heuvel, (Cambridge University Press 1983).
10. S.A. Rappaport and P.C. Joss, in *Accretion-Driven Stellar X-ray Sources*, ed. W.H.G. Lewin and E.P.J. van den Heuvel, (Cambridge University Press 1983).
11. A. Treves, L. Maraschi, and M. Abramowicz, *Publ. Astron. Soc. Pac.* **100**, 427 (1988).
12. D.L. McKenzie, *et al.*, *Ap. J.* **241**, 409 (1980).
13. C.B. Tarter, W. Tucker, and E.E. Salpeter, *Ap. J.* **156**, 943 (1969).
14. T.R. Kallman and R. McCray, *Ap. J. Suppl.* **50**, 263 (1982).
15. M. Arnaud and R. Rothenflug, *Astr. Ap. Suppl.* **60**, 425 (1985).
16. M. Klapisch, *Comp. Phys. Comm.* **2**, 239 (1971).
17. A. Bar-Shalom, M. Klapisch, and J. Oreg, *Phys. Rev. A* **38**, 1773 (1988).
18. E.B. Saloman, J.H. Hubble, and J.H. Scofield, *Atomic Data Nucl. Tables* **38**, 1 (1988).
19. J.H. Scofield, *priv. comm.* (1988).
20. D.A. Liedahl, S.M. Kahn, A.L. Osterheld, and W.H. Goldstein, *Ap. J.* (Letters) **350**, L37 (1990).
21. D.A. Liedahl, S.M. Kahn, A.L. Osterheld, and W.H. Goldstein, *in preparation*.
22. H.E. Mason, A.K. Bhatia, S.O. Kastner, W.M. Neupert, M. Swartz, *Solar Physics* **92**, 199 (1984).
23. A. Gabriel and C. Jordan, *M.N.R.A.S.* **145**, 241 (1969).
24. A.K. Pradhan and J.M. Shull, *Ap. J.* **249**, 821 (1981).

Author Index

A

Angert, N., 15
Antonetti, A., 58
Asmussen, K., 131
Audebert, P., 58

B

Barnsley, R., 131
Bar-Shalom, A., 68
Becker, R., 15
Beirsdorfer, P., 26
Benattar, R., 58
Bennett, C. L., 26
Bernstein, E. M., 15
Breger, P., 111

C

Chambaret, J. P., 58
Chen, M. H., 26
Chu, C. C., 131

D

de Heer, F. J., 111
DeWitt, D. R., 26
Djurović, S., 157
Dunn, G. H., 15

E

Ebeling, W., 97

F

Feldman, U., 171
Fields, D. F., 78
Frank, A., 15

Frieling, J., 111
Fußmann, G., 131

G

Gauthier, J. C., 58
Geindre, J. P., 58
Goldstein, W. H., 68, 78, 181

H

Haselbauer, J., 15
Henderson, J. R., 26
Hoekstra, R., 111
Horton, L. D., 111

J

Janeschitz, G., 131
Jennewein, E., 15

K

Kahn, S. M., 181
Kleinod, M., 15
Knapp, D. A., 26

L

Lee, R. W., 39
Leike, I., 97
Leonhardt, U., 97
Levine, M. A., 26
Liedahl, D. A., 181

M

Mandelbaum, P., 86
Mandl, W., 111

Marrs, R. E., 26
McCracken, G. M., 144
Mokler, P. H., 15
Morsi, H., 111
Müller, A., 15
Mysyrowkz, A., 58

N

Neumann, J., 15

O

Olthoff, J. K., 157
Oreg, J., 68
Osterheld, A. L., 181

P

Perry, T. S., 78
Phaneuf, R. A., 3, 15
Pracht, U., 15
Pröbstel, U., 15

R

Roberts, J. R., 157

S

Salzborn, E., 15
Samm, U., 144

Schennach, S., 15
Schneider, D., 26
Schneider, M. B., 26
Schumacher, U., 131
Scofield, J. H., 26
Sobolewski, M. A., 157
Spies, W., 15
Springer, P. T., 78
Stenke, M., 15
Stewart, R. E., 78
Summers, H. P., 111

U

Uwira, O., 15

V

Van Brunt, R. J., 157
Völpel, R., 15
von Hellerman, M., 111

W

Wagner, M., 15
Whetstone, J. R., 157
Wilson, B. G., 78
Wolf, R., 111
Wróblewski, D., 121

7